AF749309

Application of Optical Fiber in Engineering

Edited by Sulaiman Wadi Harun

Published in London, United Kingdom

IntechOpen

Supporting open minds since 2005

Application of Optical Fiber in Engineering
http://dx.doi.org/10.5772/intechopen.91601
Edited by Sulaiman Wadi Harun

Contributors
Norshamsuri Ali, Syed Alwee Aljunid Syed Junid, Rosdisham Endut, Nor Roshidah Yusof, Mohd Rashidi Che Beson, Rui Min, Husam Abduldaem Abduldaem Mohammed, Mohd Hanif Hanif Yaacob, Lelio De La Cruz-May, Efraín Mejía Beltrán, Olena Benavides, Rafael Sanchez Lara, Vjaceslavs Bobrovs, Sandis Spolitis, Jurgis Porins, Janis Braunfelds, Toto Saktioto, Velia Veriyanti, Sopya Erlinda, Yoli Zairmi

© The Editor(s) and the Author(s) 2021
The rights of the editor(s) and the author(s) have been asserted in accordance with the Copyright, Designs and Patents Act 1988. All rights to the book as a whole are reserved by INTECHOPEN LIMITED. The book as a whole (compilation) cannot be reproduced, distributed or used for commercial or non-commercial purposes without INTECHOPEN LIMITED's written permission. Enquiries concerning the use of the book should be directed to INTECHOPEN LIMITED rights and permissions department (permissions@intechopen.com).
Violations are liable to prosecution under the governing Copyright Law.

(cc) BY

Individual chapters of this publication are distributed under the terms of the Creative Commons Attribution 3.0 Unported License which permits commercial use, distribution and reproduction of the individual chapters, provided the original author(s) and source publication are appropriately acknowledged. If so indicated, certain images may not be included under the Creative Commons license. In such cases users will need to obtain permission from the license holder to reproduce the material. More details and guidelines concerning content reuse and adaptation can be found at http://www.intechopen.com/copyright-policy.html.

Notice
Statements and opinions expressed in the chapters are these of the individual contributors and not necessarily those of the editors or publisher. No responsibility is accepted for the accuracy of information contained in the published chapters. The publisher assumes no responsibility for any damage or injury to persons or property arising out of the use of any materials, instructions, methods or ideas contained in the book.

First published in London, United Kingdom, 2021 by IntechOpen
IntechOpen is the global imprint of INTECHOPEN LIMITED, registered in England and Wales, registration number: 11086078, 5 Princes Gate Court, London, SW7 2QJ, United Kingdom
Printed in Croatia

British Library Cataloguing-in-Publication Data
A catalogue record for this book is available from the British Library

Additional hard and PDF copies can be obtained from orders@intechopen.com

Application of Optical Fiber in Engineering
Edited by Sulaiman Wadi Harun
p. cm.
Print ISBN 978-1-83962-614-2
Online ISBN 978-1-83962-615-9
eBook (PDF) ISBN 978-1-83962-619-7

For EU product safety concerns:
IN TECH d.o.o., Prolaz Marije Krucifikse Kozulić 3, 51000 Rijeka, Croatia, info@intechopen.com or visit our website at intechopen.com.

We are IntechOpen, the world's leading publisher of Open Access books

Built by scientists, for scientists

5,300+
Open access books available

130,000+
International authors and editors

155M+
Downloads

156
Countries delivered to

Our authors are among the
Top 1%
most cited scientists

12.2%
Contributors from top 500 universities

CLARIVATE ANALYTICS BOOK CITATION INDEX INDEXED

WEB OF SCIENCE™

Selection of our books indexed in the Book Citation Index in Web of Science™ Core Collection (BKCI)

Interested in publishing with us?
Contact book.department@intechopen.com

Numbers displayed above are based on latest data collected.
For more information visit www.intechopen.com

Meet the editor

Sulaiman Wadi Harun received his bachelor's degree in Electrical and Electronics System Engineering from Nagaoka University of Technology, Japan, in 1996, and his master's and doctoral degrees in Photonics Technology from the University of Malaya in 2001 and 2004, respectively. He has nearly 20 years of research experience in the development of optical fiber devices including fiber amplifiers, fiber lasers, and fiber optic sensors. Professor Harun has published more than 900 articles in reputable ISI journals, and his papers have been cited more than 8000 times with an h-index of 40, showing the impact on the community. He is the Fellow of the Malaysian Academic of Science, and the founder and honorary advisor for the Optical Society of Malaysia.

Contents

Preface

The rapid developments in optical communication and integrated optics have expanded the interest in guided-wave optics, in which optical waveguide and optical fiber play a central role. The guided-wave photonics technology now plays a role in generating and processing information for sensing and other optical device applications in addition to its original application of transmitting information in fiber optic communication. The interaction between the propagating light with the optical fiber material is the foundation of the development of various applications such as optical amplifiers, fiber lasers, sensors, etc. The use and demand for optical fibers have grown in tandem with numerous new applications and this book aims to present the latest research in the field of optical fiber technology.

Chapter 1 investigates the birefringence and polarization mode dispersion characteristics in various commercial single-mode fibers. It is shown that the frequency value has an important role in determining the effectiveness of the optical fiber performance and stability of power delivery applications. Chapter 2 describes the properties of a quantum signal in optical fiber and its application for quantum communication.

Optical fiber sensors, another important application of optical fiber, have also experienced rapid development and attracted tremendous interest in recent years for various practical applications. Compared with the conventional electrical or other types of sensors, fiber sensors have many advantages including their immunity toward electromagnetic interference and their ability to be deployed in a corrosive and volatile environment. Optical fiber can not only transport information acquired by sensors at a high speed and in large volume but can also play the role of a sensing element itself. For instance, fiber Bragg grating (FBG) sensors can be used for monitoring temperature, strain, water level, humidity, etc. Grating devices in polymer optical fiber (POF) have attracted interest in recent years due to their potential applications in various fields, especially biomedical. Chapter 3 presents the state of the art about fabrication technology of grating devices in different kinds of POFs and explores potential sensing application scenarios, and it focuses on the fabrication of chirped POF-FBG devices and the potential application of such devices.

Chapter 4 shares the development of several optical fiber platforms for an ammonia gas–sensing application based on modifying an optical fiber cladding layer through an etching and tapering process. This modification extends the evanescent wave to propagate outside of the optical fiber and thus enhances its interaction with the analyte. Chapter 5 studies the integration of FBG sensors into a single-channel and a multichannel spectrum-sliced wavelength division multiplexed passive optical network (SS-WDM-PON). The operation of both sensor and data transmission systems, over a shared optical distribution network (ODN), is also evaluated to provide stable, high-performance mixed systems in the future.

The use of stimulated Raman scattering (SRS) as a means of amplifying signals in telecommunication systems has not been widely used because of the lack of reliable high-power pump sources. Raman fiber laser (RFL) could be used for this application and thus many research works have been devoted to this area in recent years. Chapter 6 presents the analysis of energy transfer among Stokes components in RFS (Raman Fiber Scattering).

Sulaiman Wadi Harun
Department of Electrical Engineering,
University of Malaya,
Kuala Lumpur, Malaysia

Chapter 1

Simulation of Birefringence and Polarization Mode Dispersion Characteristics in Various Commercial Single Mode Fibers

Toto Saktioto, Yoli Zairmi, Sopya Erlinda and Velia Veriyanti

Abstract

Single mode optical fiber operation for long haul distance communication media has rapidly developed. Several efforts are implemented to reduce and control the attenuation and absorption of signal propagation. However, fiber parameters still experienced interference with internal and external factors that result birefringence and polarization mode dispersion such as bending power losses, signal widening and increasing wavelengths. In order to reduce and optimize the interference which is experimentally difficult to demonstrate because of the very long fibers hence a numerical simulation is set with perspective of twisted fiber disorder as a function of wavelengths and fiber geometry. The simulation evaluates the various refractive indices, radius of fibers and wavelength sources. The quality of optical fiber interference can be identified from the twisted power losses values with different variations of twisted radius. This model obtained indicates the greatest power losses occurring as a function of radius, refractive indices and wavelength. The results show that normalized frequency value has important role in determining the effectiveness the optical fiber performance and stability of power deliver. The addition of wavelength can affect the fibers experiencing birefringence and polarization mode dispersion occurring at wavelength of telecommunication regimes.

Keywords: single mode fiber, birefringence, power loss, wavelength

1. Introduction

The use of fiber optic in telecommunication circuits and networks cannot avoid the influence and appearance of waveguide geometry problems such as bending and twisting although the growth and investigation of telecommunication hardware and software components to minimize the negative effects still have rapidly developed over the last 30 years [1–3]. This is observed in the conduct of studies to obtain easier, cheaper, accurate, clear, and low-power solutions and products to transform network components with the expectation of ensuring no interruption while sending photonic signal information [4, 5]. Fiber parameters still experience interference with internal and external factors that result birefringence and polarization mode dispersion. Not only power attenuation and signal defects but also due to material factors and the effect of the power transmitted at a certain

IntechOpen

wavelength then it can reduce the signal profile and quality. However, there are scientific challenges and environmental error optical fiber products carrying waves of information observed in the waveguide medium. It is therefore important to investigate these problems from the perspective of both the material and environmental factors [6, 7].

One of the media components of telecommunications is optical fibers and a discrepancy of refractive indices from the orthogonal polarization was observed due to the changes or damages to the size of its cylindrical symmetry. This further leads to a change in its propagation speed. This phenomenon is often called birefringence and mostly leads to Differential Group Delay (DGD) between two polarized waves and this is not desirable in the long-haul telecommunication system. This performance partially can be observed in fiber bending. Moreover, the variation in this concept causes random changes in the Polarization Mode Dispersion (PMD) [8]. Birefringence can be caused by intrinsic factors such as geometric and stress as well as extrinsic ones such as laterals stress, bending, and twist. This study considered the extrinsic factors and their possible effects on the intrinsic ones in each fiber but they are kept constant one of them to determine the level of wave disturbance produced. However, the PMD parameter has been reported to influence the bit rate for optical communication systems such that when it has a small value, a higher bit rate is produced on the fiber [9]. The restrictions on the extrinsic factors were presumed to be placed in order to clearly understand the physical function of intrinsic factors such as the contribution of geometry and stress of the material to the interference of waves propagated in single-mode optical fibers and to provide commercial recommendations for fiber products to be designed and fabricated. Meanwhile, extrinsic factors are better understood from environmental disturbance and human errors on the fiber path used.

The inevitable imperfections in the manufacture of fiber optic lead to birefringence, which has been discovered to be one of the causes of signal widening in optical fiber. Such imperfections can be geometrical such as a circular or elliptical symmetry problem, since the mechanical force applied and/or bending of the fiber. Birefringence is also intentionally introduced, for example, by using the cross-section elliptical, circular in order to produce polarization-maintaining fibers.

This chapter, therefore, proposed the simulation of optical fiber with due consideration for several factors and physical quantities of controllable and readily variable parameters to mathematically and physically interfere with wave propagation. This simulation is required during the experiment due to its low cost, ability to manipulate and control several variables, and applicability in fibers production. It also aids the determination of the effect of polarization on the cylindrical fiber imperfect core as wellas to prevent the significant effect of the dispersion magnitude determined. In Single-Mode Fibers (SMF) pulses, light waves have a limited spectral in a single state with the output usually widened while the polarization spreads throughout the whole series of fibers [10]. However, it is possible to explain this state of polarization using birefringence capital as observed in the effective difference in the index for orthogonal polarized normal modes [1]. The dispersion observed is PMD due to its small value compared to the others. Therefore, it is very important to conduct further study on the development of fiber design in communication systems to ensure better performance.

Optically, power losses due to bending in decibel unit can be expressed by:

$$\mathrm{L} = 10\log_{10}\frac{P_i}{P_o} = 10\log_{10}\left[\frac{1}{\exp(-2\propto L)}\right] = 4.342(-2\propto L) \tag{1}$$

where P_i is input power (Watt), P_o is output power (Watt), α is power loss coefficient (m^{-1}), and L is length of fiber (m) [11].

Therefore, in order to minimize power losses and geometry disturbance caused by birefringence, this chapter, however, investigated the characteristics of birefringence and PMD of commercial optical fiber using the OptiFiber system. OptiFiber is optical fiber software to design and calculate several optical parameters and functions. Fiber optics may have graded index fibers and concentric layers of lossless materials that can be estimated by using a sequence of constant index layers. In term of mode solution, OptiFiber can solve an exact solution based on terms and boundary conditions at boundaries instead of relying on meshes to estimate the geometry and structure. The mode solvers are important for single mode and multimode fiber calculations as many modes in the spectrum will occur. OptiFiber allows users to decompose any field into the fiber modes. The modes complex coefficients for any field can be calculated. In the same way, by defining the intensity or the amplitude of modes, OptiFiber can demonstrate the total (composition of modes). This is able to compute the single mode and multimode field after propagating down the fiber by a specified distance. Therefore, SMF and refractive indices ones are evaluated for wavelength sources of 1310 nm and 1550 nm in normalized frequency regimes over the radius and fiber bending. Thus, it can give a better contribution for optical application purposes of fibers.

2. Theoretical aspects

Birefringence is the optical property of a material having a refractive index depending on the polarization and propagation direction of light [1]. These phenomena can be called as birefractive (or birefringent) representing optically anisotropic materials. The birefringence is often quantified as the maximum difference between refractive indices exhibited by the material such as crystals with structures of non-cubic crystal. Those can be meant as plastics under mechanical stress. It is also responsible for the phenomenon of double refraction where by a ray of electromagnetic wave, if wave incident upon a birefringent material, is split by forming polarization into two rays taking slightly different paths. Danish scientist Rasmus Bartholin in 1669, who firstly described and observed this effect [7] in calcite, a crystal which have one of the strongest birefringence. But, Augustin-Jean Fresnel that figured out the phenomenon of polarization, understanding spectral of light as a wave with field components in transverse polarization (perpendicular to the direction of the wave vector). This event was not until the 19th century. Birefringence represents double refraction occurs if electromagnetic waves are launched to a certain material waveguide and it will split into two different signals. Two kind examples of optical material causing birefringence are Boron nitride and Calcite crystals. Birefringence can splits the signals into two images of the similar color as a “double image”. It just as a prism splits waves into a large number of colors.

The original optical fiber does not have a perfect cylindrical core due to its varied diameters and this causes voltage unevenness along the fiber. It also led to the difference in the propagation constant of the two polarization components then consequently, the optical fiber becomes birefringence. Moreover, the incorporated linear polarized light caused an assumption of the same amplitude for the two polarization components with no phase difference observed at the output end. However, the propagation of the light along the fiber led to the exit of one mode in the other phase due to the propagation constant of different phases. Therefore, at each point along the fiber (for random phase differences), the two components have the ability to produce elliptically polarized light while at π/2, it is circular. This means there is the development of the polarization from linear to an ellipse to circle to ellipse and back to linear and this alternating sequence has been reported to be continuing along the fiber [11, 12].

Birefringence is caused by both intrinsic and extrinsic factors. For example, intrinsic disturbance accidentally occurs in a factory-made process to be a permanent feature of the fiber. This includes the noncircular core causing the geometric aspect and the asymmetrical fields producing the stress aspect in the fiber around the core region. The external forces found to be causing the birefringence include lateral pressure, bending, and twisted fibers during handling and cabling process. These three mechanisms are, however, usually present to some extent in telecommunications fiber [13, 14]. Birefringence is the difference between the polarization eigenmode propagation constants shown as [7],

$$\Delta\beta = \beta_x - \beta_y \tag{2}$$

Birefringence caused by lateral stress can be expressed by,

$$\Delta\beta_{Lateralstress} = -8\frac{Cpk_0}{\pi d}\left[1-\left(\frac{a}{d}\right)^2 H(V)\right] \tag{3}$$

While birefringence caused by bending is defined as follows,

$$\Delta\beta_{Bending} = -\frac{1}{8}\left(\frac{d}{R}\right)^2 ECk_0\left[1-\frac{1}{3}\left(\frac{a}{d}\right)^2 H(V)\right] \tag{4}$$

Birefringence caused by stress can be expressed by,

$$\Delta\beta_{Tension-coiled} = -2\frac{2-3v}{1-v}C\frac{f}{\pi dRc}k_0 \tag{5}$$

where,

$$H(V) = 2+\frac{4(U^2-W^{2)}}{U^2V^2W^2}+\frac{4}{U}\frac{J_0(U)}{J_1(U)} \tag{6}$$

$$U = a\sqrt{n_1^2k_o^2-\beta^2} \tag{7}$$

$$W = a\sqrt{\beta^2-n_2^2k_o^2} \tag{8}$$

$$V = \left(U^2-W^2\right)^{1/2} = k_0 a\left(n_1{}^2-n_2{}^2\right)^{1/2} = \frac{2\pi a}{\lambda}\left(n_1{}^2-n_2{}^2\right)^{1/2} \tag{9}$$

Where V is normalized frequency, β = propagation constant, C = Photo-elastic constant, p = lateral force, k_o = wave propagation constant in vacuum, E = Young modulus, a = core radius, d = the outer diameter of the fiber, f = axial tension, c = speed of light in vacuum, υ = Poisson's ratio, n_1 = core refractive index, n_2 = cladding refractive index, and λ = wavelength.

Light is linearly polarized through SMF assuming that the two polarization components have similar amplitude and at the output end there is no difference of phase. But when light propagates along the fiber, one mode exits from difference phase due to propagation constants. Thus, at each point along the fiber (for random phase differences) the two components will produce an elliptical polarized light. In phase difference $\pi/2$, a circular polarized light can be produced. In this case, the changes of polarization are from linear to ellipse to circle to ellipse and back to linear. This alternating polarization sequence continues along the fiber. The L_p length of the polarizing fiber rotated through 2π radians angle is expressed as the length of the fiber pressure given by the equation as follows,

$$L_p = \frac{2\pi}{\delta\beta} \tag{10}$$

The widening of the pulse in SMF is caused by birefringence. This happens when the input pulse moves the two orthogonal polarized components of the basic fiber mode at different group speeds and group velocities of V_{gx} and V_{gy}, to arrive at the ends of the fiber with length z. The delayed time, ΔT, between the two orthogonal polarized components is calculated by [12].

$$\Delta T = \left| \frac{z}{V_{gx}} - \frac{z}{V_{gy}} \right| \tag{11}$$

This difference in propagation time leads to an expansion of pulses called PMD which is a limiting factor especially in long-distance optical fiber communication systems operating at high bit rates. However, assuming the fibers have a constant birefringence, it applies only to those maintaining polarization.

PMD is one of modal dispersion representing a distortion mechanism happening in multimode fibers and other waveguides, where the signal is distributed in time since the traveling velocity of the optical light is not similar to all modes. Other names for these phenomena include kinds of distortion and dispersion such as multimode distortion, multimode dispersion, modal distortion, intermodal distortion, intermodal dispersion, and intermodal delay distortion. Modal dispersion in a step-index fiber optics can be analogically compared with multipath propagation of a radio signal. Spectra of optical wave launch to the fiber with various angles to the fiber axis, until the fiber's acceptance angle. Waves transmit to a shallower angle travel by a more direct path, and arrive sooner than optical spectra those enter at a steeper angle (that reflects many times off the boundaries of the core as they propagate to the length of optical fiber). The arrival of different components of the signal at different times distorts the shape [3]. Modal dispersion limits the bandwidth of multimode fibers. For instance, a typical step-index fiber with a 50 μm core would be limited to approximately 20 MHz for a one kilometer length, then it can be said as a bandwidth of 20 MHz km. Modal dispersion can be considerably reduced, however it never completely eliminated, by the use of a core which have a graded refractive index profile. But, multimode graded-index fibers that have bandwidths exceeding 3.5GHz·km at 850 nm are nowadays commonly manufactured for use in 10Gbit/s data links.

Modal dispersion is different from chromatic dispersion, a distortion that will produce since the differences in propagation speed and direction of various wavelengths of optical spectra. Ideally, modal dispersion can occur, as monochromatic electromagnetic wave. A particular case of it is polarization mode dispersion (PMD), a dispersion phenomenon of fiber generally associated with single-mode fibers. This PMD results if two modes that normally propagate at the similar speed because of fiber core geometric and stress symmetry (for instance, two orthogonal polarizations in a waveguide of circular or square cross-section), launch at different speeds because of random imperfections that break the symmetry.

PMD has linearly effect when electromagnetic wave propagates in a resonator waveguide, namely "single-mode" fibers. Although the waveguide as a single mode, these optical fibers contribute two modes of traveling pattern distinguished by their polarization. Due to the fiber experiences the birefringence, the two modes propagate with different group velocities, and this birefringence random change along the length of fiber produces in arbitrary coupling between the modes. PMD phenomena lead to pulse distortion and system impairments in term of current practical transmission technology, resulting that limit the transmission capacity of the fiber. This study has been reviewed by researcher [3, 13], generally covering the demonstrative aspects and applications of PMD characteristics to fiber waveguide systems and the PMD effects on nonlinear fiber waveguide. In this study, it is aimed to complement the investigation and to analyze and synthesize the basic principle and concept of PMD, interweaving and linking the basic rule and mathematical formulation that emerge scattered in kind sources in the references. Author will describe the relationship between time and frequency domain analyses and the isomorphic connection between the three-dimensional (3-D) by using real-valued 3-D Stokes vectors and the two-dimensional (2-D) view using complex-valued 2-D Jones vectors. Isomorphic pairings of users such as these have been greatly applied elsewhere in photonics and optoelectronics such as in opto-mechanics, in opto-quantum mechanics [7], and also in the unification of quantum theory and general relativity [8]. Authors consider this methodology for the objective and aims.

Basically at least there are three different dispersive characteristics in fiber optics of which dispersion can occur in polarization effect and is the most complex. One of fiber mode systems, i.e., a multimode fiber system, a signal pulse can separate into multiple spatial paths or modes [13]. Each mode or component may arrive at the receiver at a certain time, widening the received signal. Another mode fiber, single-mode fiber answers the differential mode delay cases, will allow data values to be gone up until chromatic dispersion — the variation of propagation speed over the wavelength — results unacceptable signal broadening [12, 14]. The quantity of chromatic dispersion that a system can influence is inversely proportional to the square of the bit rate since an increment data values means not only narrower bit slots, but also a wider spectrum and increased spreading that are more sensitive to the widening of neighboring signals [15].

In polarization concept, namely The Principal States of Polarization model [Poole and Wagner, 1986] [15–17] is referred to the experiences that at any given optical parameters such as frequency, there exists a set of two mutually orthogonal input states of polarization effect where the relationship output states of polarization phenomena are found that it independent of frequency to first order. The Differential Group Delay (DGD) producing from PMD is then introduced for the two output principal PSPs. Generally, in telecommunication single-mode fibers system the birefringence varies randomly along the length of fiber, an artifact of variation in the drawing and cabling process. In addition, involving to the thermodynamic function and dependence of the disturbances that deal with on the fiber, the propagation properties of wave particularly will vary with ambient temperature and circumferences [16]. Practically, fluctuations in thermodynamic function such as temperature strongly influence PMD time evolution. To resolve properties of long fiber spans, one considers a statistical mechanics approach. Therefore, in this situation of long span fibers, the effects of polarization Eigenstates can only be introduced locally and the birefringence vector has to be considered as stochastic phenomena.

In fiber optics, generally there are slightly several discrepancies in the traveling characteristics of electromagnetic waves with different sates of polarization. A *differential group delay* can happen even for optical fibers which in accordance with the design will have a rotational symmetry and therefore they exhibit without birefringence. This case is able to produce random imperfections such as bending, twisting of the fiber optics, or from other kinds of mechanical perturbations stress, strain as the results of mechanical and temperature changes. Especially, because of the effect of influence of twisting and bending, the cabled fiber having PMD can be completely different from the similar to optical fiber on a spool. However, nowadays new designs can be used to optimize and apply fiber-optic links in order to reduce the disturbances or low PMD, but the mobile and handling of such cables is still able to have several disturbances. The terms PMD and DGD are misconducted in solving the dispersion or they are interchangeably often used. However, they sometimes have slightly different meanings. Several researchers use the phenomena of PMD and consider DGD to be its magnitude. Others definition of PMD is as the statistical deviation formulation of DGD in several wavelength range. It can be noted that for fiber optics, actually the DGD can have a complicated and substantial relationship over fiber optic wavelength source. Nevertheless, other researchers apply the *second-order* term of PMD for the derivative of the differential group delay in term of the angular frequency, even though there is actually no second-order derivative involved. PMD can contribute to pulse dispersion and therefore causes transmission impairment. Although PMD is usually thought of as a fiber transmission effect, it is supported by the repeater has to be taken into account since the sum of PMD in a trajectory is made up of contributions from the repeater and the transmission fiber. The objective of the repeater design is to produce the PMD contribution of the repeater smaller than optical fiber (that is <0.15 ps/km). The largest potential components of PMD are erbium-doped fiber and the isolator. Isolator components are designed and modified to have a PMD compensation element which they greatly can reduce the PMD from that found in standard single-stage devices (generally more than 1 ps). PMD is also greatly reduced by dual-stage isolators. One has to care during the manufacture of the erbium-doped fiber in order to minimize its PMD. A repeater's PMD on average can be expected to be less than 0.30 ps. On the other hand, the repeater's PMD cannot be significantly contributed by the couplers.

PMD, as a modal dispersion with two kinds polarizations of signal in the fiber, generally traveling with a similar speed, propagate at various speeds due to random asymmetries and, imperfections hence it results any distribution of signals. If it is not compensated, that is not easy; it ultimately limits the value where signals can be launched over a fiber. Moreover, in an ideal fiber, the geometry core has a perfectly circular and, the fundamental mode has two orthogonal polarizations traveling at the same speed. Also, the signal is randomly polarized through ahaphazard superposition of the two polarizations, however; since it is in an ideal situation, an identical degeneration of the polarization occurs. However, in a commercial fiber, there are uncertain imperfections having damage the cross section symmetry and causing the propagation of the polarization at different speeds. In this case, the components of a signal slowly separate and this, for example, causes the signals to propagate and overlap. Due to the randomness of the imperfections, the signal spreading effects in SMF correspond to a random traveling, and hence have a mean polarization-dependent time-differential ΔT which is also introduced as the Differential Group Delay (DGD) proportional to the square root of propagation distance L. Therefore, the PMD-induced pulse widening estimates are made using the following relationship [16, 17]:

$$\Delta T = D_{PMD}\sqrt{L} \tag{12}$$

For long SMF, PMD values are calculated in the form of average DGD values using the following equation [10],

$$\langle \Delta\tau \rangle = \sqrt{\frac{8}{3\pi}}\Delta\beta'\sqrt{l_c}\sqrt{z} \tag{13}$$

PMD can be also be calculated as a root mean square, (RMS),

$$\sqrt{\langle \Delta\tau^2 \rangle} = \Delta\beta'\sqrt{l_c}\sqrt{z} \tag{14}$$

where T = total time delay, D = dispersion parameter, z = length of fiber, l_c = length of coupling, τ = time delay.

3. Single mode fiber: performance assessment

The SMF parameters input numerically demonstrated by OptiFiber is shown in **Table 1**. The simulations were conducted to determine the SMF birefringence and PMD profile using the core and cladding parameters of each fiber with core diameter and cladding kept constant at 4.1 μm and 62.5 μm respectively. Moreover, the normalized frequency was maintained while the core and cladding refractive indices differentiating the fibers are presented in **Table 1**. The value of SMFs having the range of normalized frequency of LP_{01} can be defined as core and cladding refractive indices and control single mode performance as depicted in **Figure 1**. However, the discrepancies of each sample show how power is delivered to the fiber.

Fiber optic	Core (n)	Cladding (n)
SMF 28	1.45213	1.44692
SMF 28e	1.4677	1.4624
SMF 28e+	1.45173	1.44602
SMF 28e + LL	1.45223	1.44702
SMF 28 ULL	1.44525	1.44002

Table 1.
SMF fiber optic refractive indices.

DOI: http://dx.doi.org/10.5772/intechopen.94127

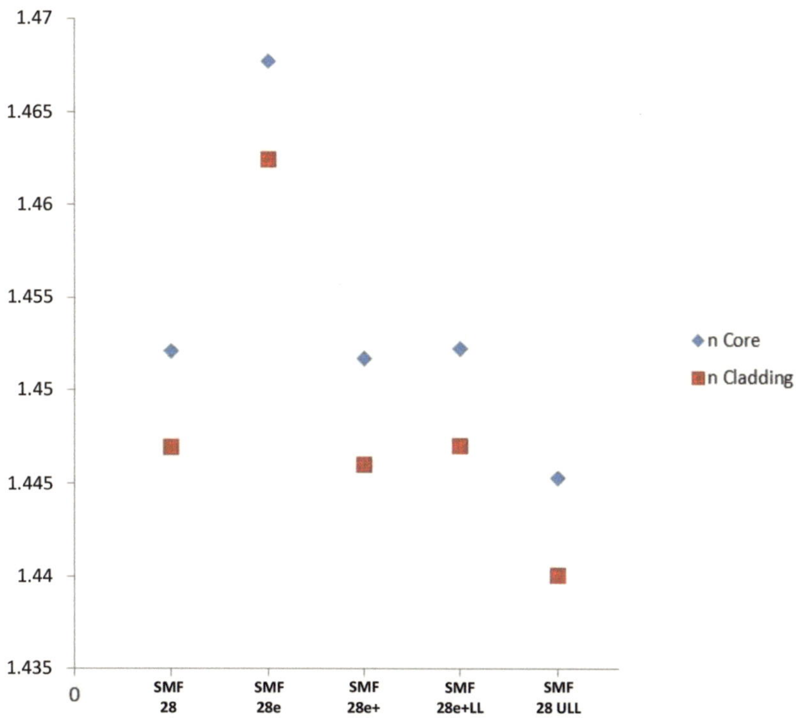

Figure 1.
*Core and cladding profile for various SMFs of **Table 1**.*

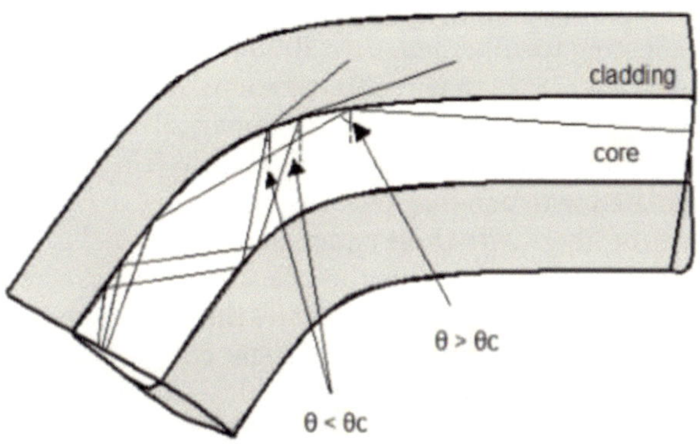

Figure 2.
Schematic profile of refractive ray due to fiber bending.

The SMF profile was determined using the Refractive Index type Profile with regions 0 and 1 which served as the core and cladding parameters of optical fiber and pure Silica while Germanium material was used as positive dopants and Florin as negative. Moreover, the optical fiber mode used to produce an index capital at a given wavelength and to determine the fiber field capital was LP mode (Matrix Method) with cutoff wavelength parameter indicated in the LP_{01} and LP_{11}. In addition, the fundamental property mode simulation was also set to determinethe default values of the material, bending, and loss parameters as illustrated in **Figure 2**. Meanwhile, in the scan section, the wavelength was adjusted by a fixed option and the values used for the part of the parameters were 1.2 to 1.6 with 100 iterations.

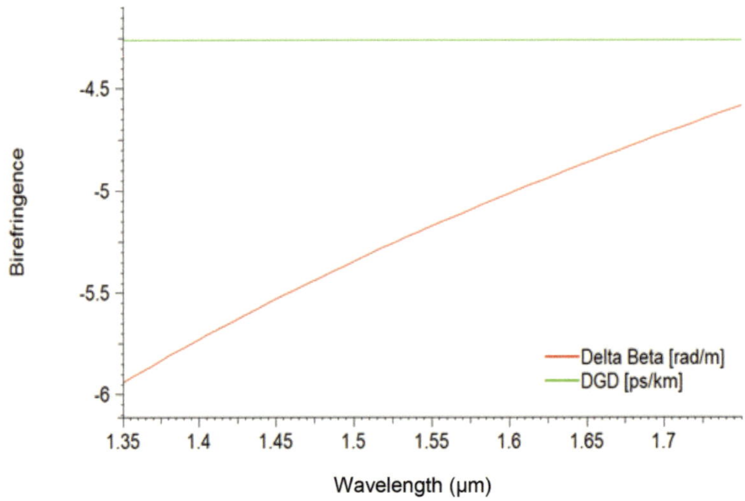

Figure 3.
Birefringence profile of SMF 28.

The birefringence caused by parameter disturbances started with the determination of the photo-elastic constant of the fiber and was found to be of 3.44 × 10^{11} m^2/kg.W, the Young modulus value of 7.75 × 10^9kg.W/m^2, and the Poisson ratio of 0.164 extrinsic factors even though it was not counted as a dominant factor. Moreover, bending and stress in the fiber were also observed to have effects with the bending discovered to be 0.12 m with a rolled fiber tension force of 0.5 N. At the output section, the value of 0.4 μm spectral range with 51 iterations was used while the PMD was obtained by adjusting the fiber length to 1000 m, the coupling length by 20 m, and the spectral length was 0.1 μm with 201 iterations.

As bending induced parameters of SMF, the numerical demonstration can explain several losses over wavelength and radii. The birefringence caused by extrinsic factors was simulated with bending and tension force of the circular fiber kept constant on all types of fibers with these parameters considered to have the same disturbance of all the samples to evaluate the effect of intrinsic factors. The results at SMF 28, 28e, 28e+, 28e + LL and 28 ULL are nearly the same as shown in **Figure 3**.

Figure 4 shows of SMF 28 profile, which is slight change in all SMF curves and by describing the discrepancies using factor 10^{-3}. The birefringence value was observed to be increasing with the wavelength (as the photon energy decreases) due to the difference in the second phase of the polarized wave while the DGD was discovered to be constant. The magnitude of birefringence at the wavelength of 1550 nm fiber SMF 28 was −5.1753668 rad/m, SMF 28e had −5.17534 rad/m, SMF 28e + had −5.17539 rad/m, SMF 28e+ had −5.14879 rad/m, and SMF 28e+ had −5.175397 rad/m. In addition, at SMF 28 ULL, the value was recorded to be greater than others and this means there was a large power reduction at this optical fiber output. It is important to note that the SMF polarized light was contributed by the magnetic and electric field. Furthermore, a greater value of birefringence or the difference in wave propagation constant was found to be causing more polarization in the optical fibers and this further led to a greater phase difference between the magnetic and electric field of light. Hence, the core is imperfectly shaped in a circle due to the bending and stress force when the fiber is rolled.

As depicted in **Figures 5** and **6**, power losses are mostly affected by 1550 nm where this wavelength has low energy. Also it readily produces power losses even SMF 28 ULL is significant amount as a result of its normalized frequency.

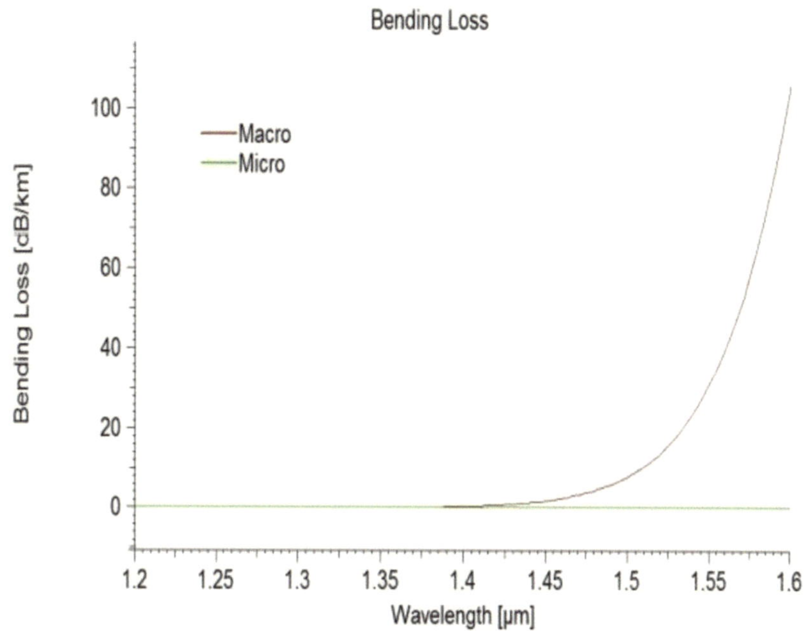

Figure 4.
Bending loss of SMF-28.

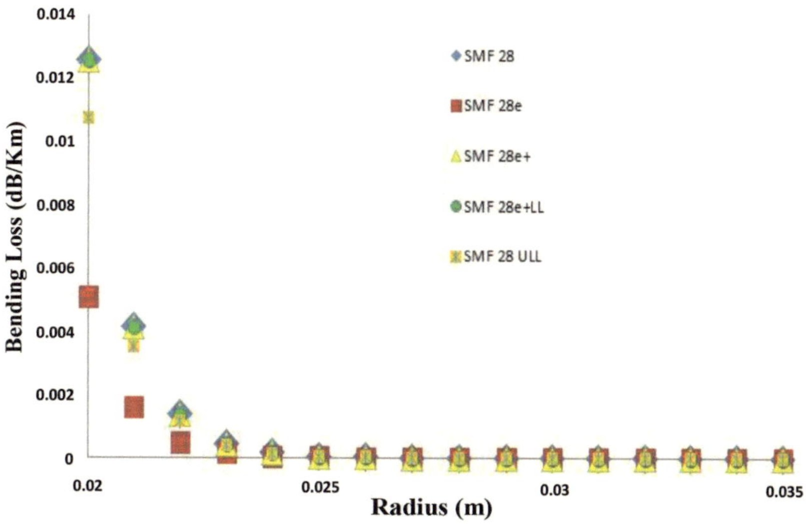

Figure 5.
Power losses due to bending of 1310 nm.

Figure 7 describes that birefringence of each fiber is nearly the same value, but they are not if corrected by 10^{-2}. They are decreased as a function of wavelength since the lower energy of wavelength source is applied, the smaller the birefringence is produced.

The simulation showed the extrinsic parameters of birefringence used in the same fiber produced different values due to the variations in the modes of each fiber and core as well as the cladding refractive indices as shown in **Table 2**.

These perturbations were accidentally known in made factory process and later become a permanent feature ofthe fiber. Moreover, a noncircular core was found to

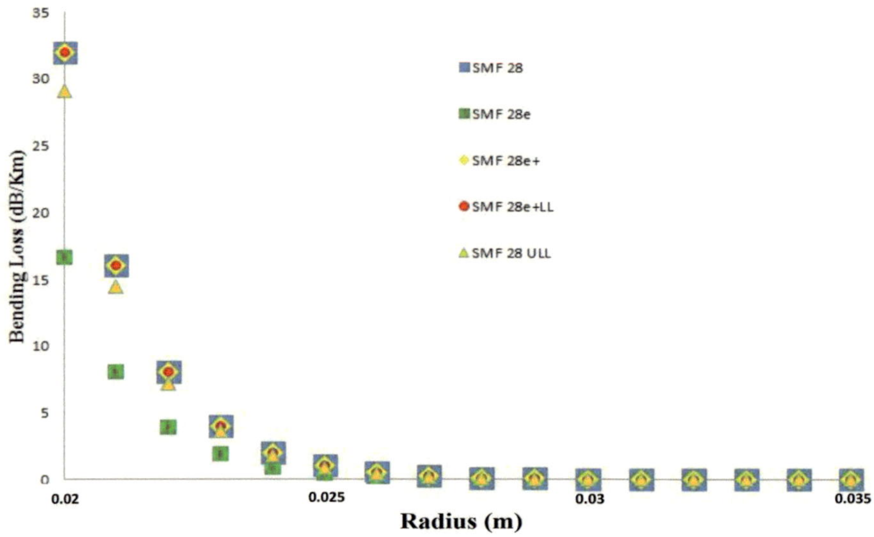

Figure 6.
Power loss due to bending of 1550 nm.

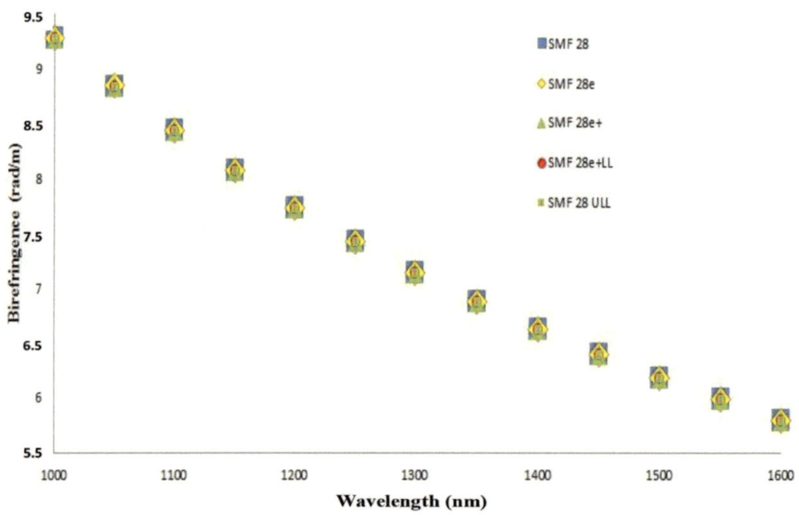

Figure 7.
Birefringences of several SMF.

Fiber optic	Birefringence (rad/m)	DGD (ps/km)
SMF 28	−5.1753668	−4.2590604
SMF 28e	−5.17534	−4.25504
SMF 28e+	−5.17539	−4.25901
SMF 28e + LL	−5.14879	−4.25906
SMF 28 ULL	−5.175397	−4.25906

Table 2.
Birefringence and DGD values.

have produced a geometric birefringence while the nonsymmetrical field caused stress birefringence. The refractive index of isotropic fibers depended on the polarization and propagation direction of light and the maximum difference between these indices was exhibited by non-cubic crystal structures. In addition, their phenomena have double refraction divided into two rays with slightly different paths by polarization. The curves representing anisotropic fibers generally refract a single incoming ray in two directions and these correspond to the two different polarizations, uniaxial or biaxial fiber. In the uniaxial one, the ray behaves according to the normal law of refraction with correspondence to the ordinary refractive index and this further makes the incoming ray to be normal at both incidence and refracting surface. However, as previously explained, the other polarization deviated from normal incidence and this means impossible to describe it using the law of refraction. In this case, the polarization components are perpendicular or ordinary and not perpendicular or extraordinary to the optic axis respectively, even in situations without double refractions.

The fiber with a single direction or optic axis of symmetry in its optical behavior was also observed to be symmetrical to the index ellipsoid, a spheroid in this case and was explained based on the refractive indices, n_α, n_β and n_γ, along Cartesian coordinate. But, two of these were known to be the same, hence, if $n_\alpha = n_\beta$ corresponding to the x and y axes, the extraordinary index is n_γ over the z-axis. The signal consists of two polarization components generally governed by different effective refractive indices, and the material with the higher was discovered to have a slower phase velocity whole the other with the lower value was the *fast ray*.

As depicted in **Table 2**, the birefringence is positive when the extraordinary index of refraction n_e is greater than the ordinary index n_o while a negative value shows that $\Delta n = n_e - n_o$ is less than zero. This, therefore, means the polarization of the slow or fast wave is perpendicularto the optic axis at a negative or positive birefringence respectively.

The circular cores did not maintain a polarization input state for more than a few meters, and this means they are not perfectly circular. Moreover, the PMD value of the single-mode optical fiber was caused by the birefringence of the fiber and this means a variation in this factor led to the random changes in the PMD [12, 17] based on the difference in the mode field diameter (MFD) of each fiber. In **Figure 8**, the PMD value fluctuated due to the variation in the birefringence value along the fiber with the wavelength. In addition, the polarization of the fiber was also discovered to have caused its dispersion while the difference in the birefringence was caused by the imperfections of the fiber core. Therefore, a greater value of birefringence led to more significant polarization as well as a great delay in the polarized wave at the output.

Table 3 describes the values of DGD and RMS for first and second-order dispersion. In the frequency domain, PMD caused the output of polarization vary with frequency for a fixed input one in a cyclic fashion. Moreover, on the Poincare sphere display, the output polarization propagates in a circular pattern on the surface where the frequency was varied. In addition, in the spectral simulation, a set of concatenated fiber trunk was randomly generated while the PMD was calculated over a range of wavelengths and DGD evaluated based on a stochastic fiber model (first-order PMD) to quantify the first order of PMD. However, the second-order showed the expression of first frequency derivative of the dispersion vector as a function of frequency and position. This means the first order curves explained more fluctuations as observed in the reduction in the PMD as the wavelength increases, but, the second-order curves are more slightly fluctuated over the wavelength and was recorded to have produced trends nearly constant compared to the first order.

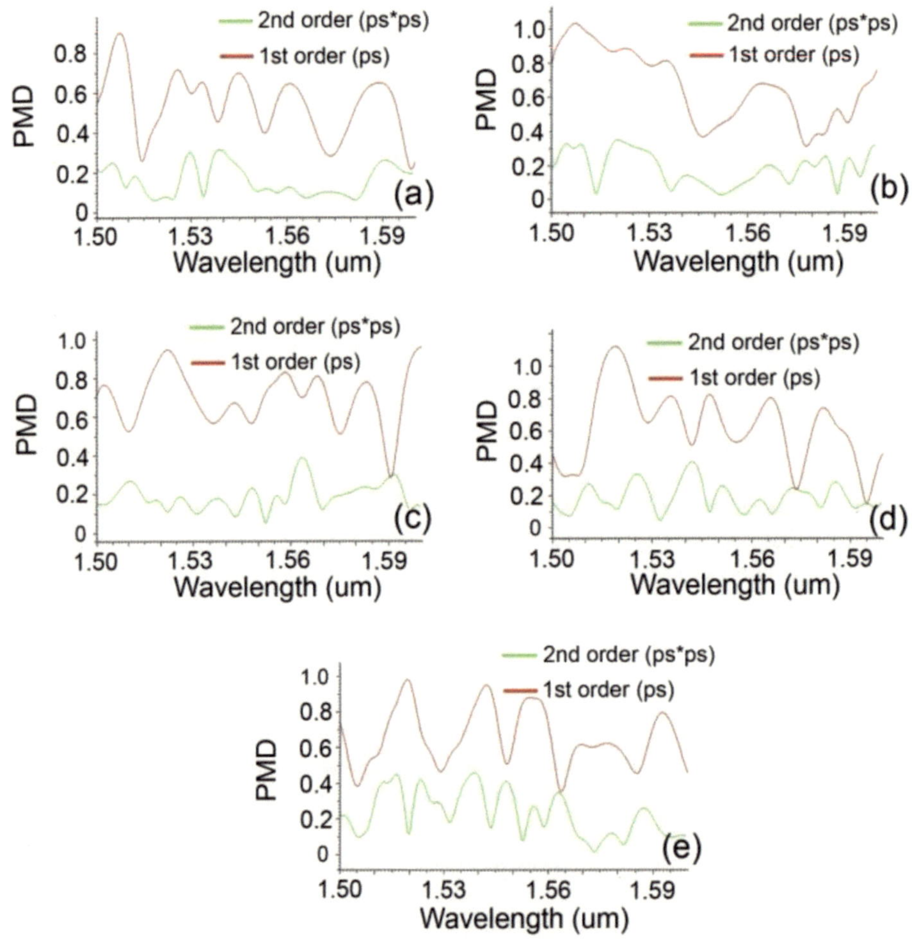

Figure 8.
PMD SMF: (a) 28, (b) 28e, (c) 28e+, (d) 28e + LL, and (e) 28 ULL.

Fiber optic	Average of DGD (ps)		RMS (ps)	
	First order	Second order	First order	Second order
SMF 28	0.6977	0.1893	0.7106	0.2003
SMF 28e	0.6630	0.1835	0.6940	0.2054
SMF 28e+	0.6977	0.1893	0.7106	0.2003
SMF 28e + LL	0.6247	0.1889	0.6639	0.2053
SMF 28 ULL	0.6523	0.2288	0.6706	0.2586

Table 3.
Average of DGD and root mean square (RMS) values.

The symmetry-breaking random imperfections were classified a geometric asymmetry as observed in the slightly elliptical cores or stress-induced fiber birefringence, in which the refractive index itself is a function of polarization. Both effects can stem from either imperfection in made factory process or from mechanical and thermal stresses imposed on thefiber in the field, moreover the latter stresses generally vary over time. A related effect is a polarization-dependent loss (PDL) that involves two polarizations suffering different rates of loss due to asymmetries. This factor similarly degraded the pulse quality. It is important to note that a circular core is not required to have two degenerate polarization states but there is a need a symmetry group which admits a two-dimensional irreducible representation. For instance in fundamental mode, an equilateral-triangle or square corehas two equal-speed polarization solutions and these common shapes also arise in crystal waveguides. However, any imperfections that make damage of the symmetry have the ability to cause PMD in such a waveguide.

The PMD has random and time-dependent effects; therefore, there is a need for an active device to respond to feedback over time. Such systems are not low cost and complex combined with the real PMD is not of the most commonly used in a certain factor in the lower data rates. Therefore, PMD-compensation systems have not been widely deployed in large scale telecommunications systems. The fiber output was essentially separated into two principal polarizations (no first-order variation of time-delay with frequency), and a differential delay was applied to re-synchronize them. Therefore, currently fibers have practical problems economically such as higher costs and losses. This means that a single-polarization fiber is required where only a single state is allowed to transmit along the fiber while the others escape because they are not guided.

4. Summary

The occurrence of birefringence in commercially optical fibers is influenced by internal and external factors such as bending and tension forces. This, therefore, changes the cores from circular to ellipses and *vice versa* and this makes light waves experienced an elliptical or circular polarization to produce two waves in different phases. Moreover, SMF 28 ULL was discovered to have the highest birefringence value while the lowest was recorded at SMF 28e + LL. Birefringence can cause PMD. SMF 28 has the largest PMD value with a value of 0.69770237 ps compared to other optical fibers and this consequently led to a large pulse widening at cladding with a low bit rate.

Author details

Toto Saktioto*, Yoli Zairmi, Sopya Erlinda and Velia Veriyanti
Universitas Riau, Pekanbaru, Indonesia

*Address all correspondence to: saktioto@yahoo.com

IntechOpen

© 2020 The Author(s). Licensee IntechOpen. This chapter is distributed under the terms of the Creative Commons Attribution License (http://creativecommons.org/licenses/by/3.0), which permits unrestricted use, distribution, and reproduction in any medium, provided the original work is properly cited.

References

[1] Kaminow, Ivan P. "Polarization in optical fibers," *IEEE J. Quantum Electron.,* vol. QE-17, pp. 15-22, 1981.

[2] Irawan, D., Saktioto, T., Ali, J., Yupapin, P., "Design of Mach-Zehnder interferometer and ring resonator for biochemical sensing", Photonic Sensors Volume 5, Issue 1, 1 March 2015, Pages 12-18.

[3] Yupapin P, Saktioto T, Ali J. Photon trapping model within a fiber bragg grating for dynamic optical tweezers use. Microwave and Optical Technology Letters. April 2010;**52**(4):959-961

[4] Irawan D, Saktioto T, Ali J, Fadhali M. Birefringence analysis of directional fiber coupler induced by fusion and coupling parameters. Optik. September 2013;**124**(17):3063-3066

[5] Saktioto, T., Ali, J., Fadhali, M., Rahman, R.A., Zainal, J., "Modeling of coupling coefficient as a function of coupling ratio", Proceedings of SPIE - The International Society for Optical Engineering Volume 7155, 2008, Article number 71551P9th International Symposium on Laser Metrology; 30 June 2008 through 2 July 2008; Code 74498.

[6] Chowdhury DQ, Nolan DA. Perturbation model for computing optical fiber birefringence from two-dimensional refractive-index profile. Opt. Lett. 1995;**20**:1973-1975

[7] Sakai J, Kimura T. Birefringence and Polarization Characteristics of Single-Mode Optical Fibers under Elastic Deformations. IEEE Journal of Quantum Electronics. June 1981;**QE-17**(6):1041-1051

[8] Wuilpart M, Rogers AJ, Defosse Y, Mégret P, Blondel M. Measurement of the spatial distribution of birefrigence in optical fibers. IEEE Photonics Technology Letters. 2000;**13**(18):836-838

[9] Chaudhary, K., Rosalan, S., Aziz, M.S., Bohadoran, M., Ali, J., Yupapin, P.P., Bidin, N., Saktioto., "Laser-induced graphite plasma kinetic spectroscopy under different ambient pressures", Chinese Physics Letters Volume 32, Issue 4, 1 April 2015, Article number 043201.

[10] Foschini GJ, Poole CD. Statistical theory of polarization dispersion in single mode fibers. J. Lightwave Technol. 1991;**9**:1439-1456

[11] Qiu, J., Zheng, D., Zhu, K., Fang, B., & Cheng, L. 2015. "*Optical Fiber Sensor Experimental Research Based on The Theory of Bending Loss Applied to Monitoring Differential Settlement at The Earth-Rock Junction*". Journal of Sensor. 2015, 1-13.to

[12] Khare RP. *Fiber Optics and Optoelectronics*. New York: Oxford University Press; 2004

[13] Poole, C.D and J. Nagel, "Polarization effect in lightwave systems," in *Optical Fiber Telecommunications IIIA*, I. P. Kaminow and T. L. Koch, Eds. New York: Academic, 1997

[14] Tahir BA, Ali Saktioto J, Fadhali M, Rahman RA, Ahmed A. A study of FBG sensor and electrical strain gauge for strain measurements. Journal of Optoelectronics and Advanced Materials. October 2008;**10**(10):2564-2568

[15] Horvath, T., Munster, P., Vojtech, J., Velc, R., Oujezsky, V. 2018. "Simultaneous Transmission of Accurate Time, Stable Frequency, Data, and Sensor System Over One Fiber with ITU 100 GHz Grid", Optical Fiber Technology. 2018, 139-143.

[16] Miah, M.S danRahman, M.M. 2012. The Performance Analysis Fiber Optic Dispersion on OFDN-QAM System. International Journal of Advanced Computing, Engineering and Application (IJACEA). 1(1);29.

[17] A. Mecozzi, "Theory of polarization mode dispersion with linear birefringence", (2008), Optic Letters. 33 (12), 1315.

Chapter 2

Quantum Signal over Optical Fiber

Nor Roshidah Yusof, Norshamsuri Ali, Syed Alwee Aljunid Syed Junid, Mohd Rashidi Che Beson and Rosdisham Endut

Abstract

This chapter aims to address the quantum signal role and properties in optical fiber application mainly in quantum communication. It covers the general discussion on quantum bits and optical waveguiding properties. The highlight of this chapter lies in the discussion of the quantum fictitious force of anti-centrifugal force which was first reported in 2001. Under this condition, the free particle experience an attractive potential towards the rotating center of a bent waveguide structure. A lot of theoretical work has been carried out to observe this quantum phenomenon. However, no intensive experimental work has been carried out to date. With the advancement of nano-fabrication technology and quantum experimental, it provides a bright potential to observe these phenomena. Thus, we proposed a promising material of Lithium Niobate on Insulator to serve as a waveguiding platform to study this quantum effect experimentally. The discussion is extended to perceive the relation between Schrodinger and Helmholtz's equation corresponding to this effect.

Keywords: quantum anti-centrifugal force, waveguide, rib waveguide, LNOI, spatial mode

1. Introduction

Optical fiber is known as one of the passive devices that serve as the transmission medium in optical communications. This cylindrical-shaped dielectric waveguide consists of the higher refractive index inner core surrounded with slightly lower index cladding which operates based on total internal reflection phenomena. To date, the advancement in optical fiber technology led to the development of new fiber designs and components to establish multiple co-existing data channels based on light propagation over distinct transverse optical modes [1]. Apart from that, there have been tremendous research and developments in quantum information processing since Richard Feynman's seminal talk on the future of quantum computing 40 years ago. These developments have been motivated by the higher demand for higher transmission bandwidth in telecommunication infrastructure. Thus, optical fiber has played its main role in the success of this technology mainly due to its high transparency and high-bandwidth support [2]. Thus, the discussion of this chapter covers the general idea of quantum bits which has become the backbone in quantum information.

IntechOpen

Apart from that, it is found that the bending of optical waveguide has led to the signal distortion which increase the propagation losses. For the pass few years, researcher put an extra effort to observe this phenomena which is known an quantum anti-centrifugal force from a quantum physics point of view.

Therefore, the aim of this chapter lies on the discussion of poorly reviewed quantum anti-centrifugal force. Furthermore, we perceive the correlation of modern interpretation of Schrodinger equation with the classical Helmholtz equation correspond to this quantum fictitious force. The observation of this effect is extended into the real application of Lithium Niobate on Insulator rib waveguide as it found to be an extensive platfrom towards the development of quantum information technology.

2. Optical field propagation

The optical waveguide structure consists of high refractive index core, surrounded by slightly lower refractive index. It operates based on total internal reflection phenomena which allow the electromagnetic wave guiding towards core. The electromagnetic plane waves propagation in homogeneous and isotropic media such as glass and diamond are governed by Maxwell's equation:

$$\nabla \times E + \mu \frac{\partial H}{\partial t} = 0 \tag{1}$$

$$\nabla \times H - \varepsilon \frac{\partial E}{\partial t} = 0 \tag{2}$$

The permittivity tensor, ε, and permeability tensor, μ is uniform. Both of Eqs. (1) and (2) are satisfied by the following plane wave solution:

$$\Psi = Ae^{-i(\omega t - k.r)} \tag{3}$$

where A and ω denotes an amplitude and angular frequency with the wavevector, $|k| = \omega\sqrt{\mu\varepsilon}$.

However, for the anisotropic media such as Lithium niobate, calcite, and quartz, both ε and μ are directional dependent [3]. The permittivity tensor is derived by the following matrix expression:

$$\varepsilon = \varepsilon_o \begin{bmatrix} n_x^2 & 0 & 0 \\ 0 & n_y^2 & 0 \\ 0 & 0 & n_z^2 \end{bmatrix} = \begin{bmatrix} \varepsilon_x & 0 & 0 \\ 0 & \varepsilon_y & 0 \\ 0 & 0 & \varepsilon_z \end{bmatrix} \tag{4}$$

where n_x, n_y and n_z are the principal indices of refraction whereas (x, y, z) denotes the principal axis of the crystal. Generally, the anisotropic media can be divided into uniaxial and biaxial crystal. In uniaxial medium, the refractive index corresponds to two equal principle elements i.e. $n_o^2 = \varepsilon_x/\varepsilon_o = \varepsilon_y/\varepsilon_o$ is defined as an ordinary index, n_o whereas the other index for the remaining principle $(n_{eo}^2 = \varepsilon_z/\varepsilon_o)$ is known extraordinary index.

The optical mode across the waveguiding area is derived from the abovementioned Maxwell's equation and Helmholtz propagation equation with a specific boundary condition. From this, it can show the deformation of spatial modes and optical loss in optical fiber [4–6]. The Helmholtz problem can be solved analytically when it is separable concerning its variables. However, this separability does not hold for all coordinate systems [7]. Therefore, the need to choose for a suitable coordinate system and boundary conditions are crucially needed prior to solve the eigenmode problem analytically.

Another limitation with regards to geometrical structure are sharp corners, e.g. a rectangular one, which requires approximations such as the assumption of strong field confinement in the waveguide [8, 9]. In cylindrical coordinates $(r; \varphi)$ the Helmholtz equation takes the form of

$$\left(\frac{1}{r}\frac{\partial}{\partial r}+\frac{\partial^2}{\partial r^2}+\frac{1}{r^2}\frac{\partial^2}{\partial \varphi^2}+k_0^2 n^2(r)\right)\Psi(\mathrm{r},\varphi)=0 \tag{5}$$

where $n(r)$ denotes the refractive index and $\Psi(r,\varphi)=\psi(r)\Phi(\varphi)$. $\psi(r)$ represents the bending mode profile.

At the bending area, it shows a higher tendency to produce radiation loss compared to straight waveguide due to the variance of phase front velocity correspond to the speed of light. These modes are strongly guided in the area with the highest refractive index. Thus, it will shift towards the outer surface of the bent waveguide. At the transition from straight to bent waveguide mode adaptation, losses will occur. Unfortunately, these losses are exponentially increased as we reduced the radii of curvature, which led to major drawbacks in developing nano-sized optical waveguiding devices.

3. Quantum anti-centrifugal force

Among quantum fictitious forces, the quantum anticentrifugal force which appears in the cylindrical symmetry system belongs to the most fascinating. This quantum phenomenon was first reported by [10] where he predicted that a free particle of vanishing angular momentum, $m=0$ experiences an attractive potential towards the rotating centre rather than repulsive. The 2-dimension time-independent Schrödinger equation of the free particle of mass, M and energy, E along the polar coordinates r and φ is given by

$$\left\{\frac{d^2}{dr^2}+\frac{2M}{\hbar^2}[E-V_m(r)]\right\}u_m(r)=0 \tag{6}$$

$$V_m(r)=\frac{\hbar^2}{2M}\frac{m^2-1/4}{r^2} \tag{7}$$

where $u_m(r)$ and $V_m(r)$ is a radial wave function and effective potential, respectively. This potential describes the existence of repulsive or attractive force on the free particle which is proportional to the value of m^2. Under the non-vanishing angular momentum $(m\neq 0)$ condition, it tends to repulse far from the rotating center thus, create the centrifugal force. On contrary, at zero angular momentum, its binding potential towards the rotating center become stronger and contribute to the arising of quantum anti-centrifugal potential, $V_Q$which is given by

$$V_Q\equiv V_0(r)=-\frac{\hbar^2}{2M}\frac{1}{4r^2} \tag{8}$$

The force corresponding to V_Q cause a ring-shaped wave packet to spread in an asymmetric form where it spreads faster towards the origin. Consequently, the particle is localized on a band-shaped domain around the origin. On contrary,

Cirone claimed that this potential force has not existed in 3-dimensional space. However, in 2010, Dandoloff proves that this counterintuitive potential existed in 3-dimensional space on wormhole geometry [11].

In 2011, he reported his finding on quantum anti-centrifugal potential dependences over the radius of the bent rectangular waveguide as depicted in **Figure 1** [12, 13]. Based on the following line element expression, the Schrödinger equation of this Cartesian coordinate can be derived as Eq. (10).

$$dr^2 = d\xi^2 + dz^2 + (1-\kappa\xi)^2 ds^2 \tag{9}$$

$$\Delta\Psi = \frac{\partial^2\Psi}{\partial z^2} + \frac{\delta^2\Psi}{\delta\xi^2} - \frac{\kappa}{(1-\kappa\xi)}\frac{\delta\Psi}{\delta\xi} + \frac{1}{(1-\kappa\xi)^2}\frac{\delta^2\Psi}{\delta\xi^2} \tag{10}$$

where $\xi \in \left[-\frac{a}{2}, \frac{a}{2}\right]$, $\kappa = \frac{1}{R}$, $C = C(a, R)$ and $\Psi = \frac{1}{\sqrt{C}}\frac{\Phi}{1-\kappa\xi}$.

The general time-independent Schrödinger equation of $\frac{i\hbar^2}{2M}\Delta\Psi = \mathrm{E}\Psi$ is then can be expressed in (ξ, s, z) coordinate system as:

$$E\frac{\Phi}{(1-\kappa\xi)^{\frac{1}{2}}} = \frac{i\hbar^2}{2M}\left[\frac{1}{(1-\kappa\xi)^{\frac{1}{2}}}\frac{\delta^2\Phi}{\delta\xi^2} + \frac{k^2}{4}\frac{\Phi}{(1-\kappa\xi)^{\frac{5}{2}}} + \frac{\partial^2\Phi}{\partial z^2} + \frac{1}{(1-\kappa\xi)^2}\frac{\delta^2\Phi}{\delta s^2}\right] \tag{11}$$

Compared to Eq. (8), Dandoloff proposed the Bohm potential, Q corresponding to the bent waveguide which is given by

$$Q = -\frac{\hbar^2}{2M}\frac{\Delta R}{R} \tag{12}$$

where R is the radius of the waveguide. This force was assumed to affect the quantum motion on a bent waveguide. However, he found that this effect has a very weak contribution to the overall potential. Thus, the quantum anti-centrifugal potential is then derived as:

$$-\frac{\hbar^2}{2M}\frac{dV_{eff}}{d\xi|_{\xi=0}} = \frac{\hbar^2}{2M}\frac{\kappa^3}{2} = \frac{\hbar^2}{2M}\frac{1}{2R^3} \tag{13}$$

It is found that at zero angular momentum (m = 0), the anti-centrifugal reciprocally dependence on R^3. Dandaloff also concluded the similarity of the stationary Schrödinger equation and Helmholtz equation for the TE modes in a waveguide structure. Subsequently, the interference pattern in the quantum medium should have the same pattern as an electromagnetic wave in a bent waveguide.

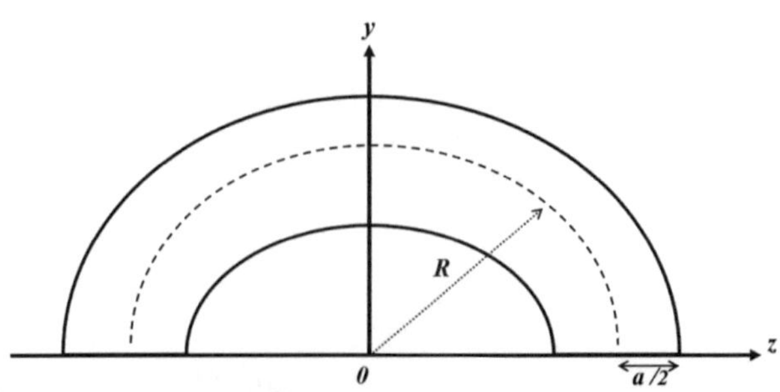

Figure 1.
Semi-circular shaped bent waveguide on Cartesian coordinates, (x,y,z).

In 2015, Victor published another finding on quantum anti-centrifugal potential by introducing half-geodesic coordinates where the curvature with constant radius corresponds to the concentric circles at $r = 0$ [7]. Its 2D Laplacian function is expressed as:

$$\Delta\psi = \frac{1}{h}\left[\frac{\partial h}{\partial r}\left(h\frac{\partial h}{\partial r}\right) + \frac{\partial}{\partial\phi}\left(\frac{1}{h}\frac{\partial h}{\partial\phi}\right)\right]\psi \tag{14}$$

In this work, an intrinsic Gaussian curvature of the surface, $K = -h^{-1}\partial_r^2 h$ and the geodesic curvature of the geodesically parallel coordinates lines $k_g = -h^{-1}\partial_r h$ has been introduced. Thus, the effective potential, V may be written as follows:

$$V = \frac{\hbar^2}{2M}\left(\frac{m^2}{h^2} - \frac{1}{4}k_g^2 - \frac{1}{2K}\right) \tag{15}$$

Within the limit of $\frac{\omega r}{c} \ll 1$, $k_g = \frac{1}{r^2} + 2\frac{\omega^2}{c^2}$ and $K = -3\frac{\omega^2}{c^2}$, Eq. (14) can be expressed as:

$$V = -\frac{\hbar^2}{2M}\left(\frac{m^2 - \frac{1}{4}}{r^2} - \left(m^2 - 1\right)\frac{\omega^2}{c^2}\right) \tag{16}$$

Eq. (15) describes the anti-centrifugal potential in a 2D plane at any surface which can be covered globally with a geodesically parallel coordinate system. At $K > 0$ (i.e elliptical), the curvature weakens the anti-centrifugal potential whereas at $K < 0$ (i.e catenoid), its potential enhanced. Recently, Dandoloff has included the quantum geometric potential (QGP) which commonly occurs at the surface of the structure, using a very high sandwich potential around the surface. The total effective potential is given by:

$$V = \frac{\hbar^2}{2M}\left(\frac{m^2}{h^2} - \frac{1}{4}k_g^2 - \frac{1}{2K}\right) \tag{17}$$

From this research, Dandoloff addressed his new finding on the total quantum effective potential which appears to be the nonlinear sigma model plus positive terms instead of the negative signage. Even though the exact number of states cannot be defined by using this Topologically Stable States, but it provides the centrifugal states additional legitimacy [14, 15].

4. Schrödinger vs. Helmholtz equation towards the measurement of quantum anti-centrifugal force

As of today, the investigation of the quantum anti-centrifugal potential was investigated purely theoretical correspond to the Schrödinger equation. However, the advancement of the nano-fabrication and quantum photonics technology since last decides has brought the impossible into reality. It has become a thriving field of research in promoting both fundamental studies of quantum phenomena and a wider range of disruptive quantum technologies. The well-developed quantum experimental setting such as the phase retrieval algorithm and single-photon detection techniques has brought the experimental on quantum anti-centrifugal potential into reality. Recently, a lot of studies on spatial mode deformation and losses in optical fiber in various geometrical structure has been carried out to observe the similar quantum effect by using the Helmholtz equation with a suitable boundary condition.

Recently, the deformation of eigenmode in the semi-circular toroidal waveguide is reported in [17] as depicted in **Figure 2(a)** by using the finite element method. The operating wavelength and refractive index of both waveguides, n_w and surrounding, n_s was set at 800 nm and 2.3 and 1.0, respectively. The mode average radial position was introduced in this simulation work in order to analyze the direction of mode. The bending radius was set at 1 μm. On top of that, the analysis was extended to a realistic structure and promising material in quantum photonic technology, Lithium Niobate on insulator (LNOI) rib waveguide [16]. Lithium Niobate (LN) is has a strong nonlinear electro-optical property which transparent over a broad spectrum spanning from visible to mid-infrared region (350 nm to 5200 nm).

Figure 2(b) depicted the rib structure which consists of 500 um thick LN substrate, 2 μm thick SiO_2 which acts as a buffer layer, and a 600 nm thin film LN. The thin film LN is etched to form the higher confinement core with height, D, and width, W of 500 nm and 1000 nm, respectively. This simulation was performed in an axial-symmetric profile at a second harmonic wavelength of 775 nm.

Based on **Figure 3(a)**, the spatial mode distorted and shifted towards the outer wall of the bent waveguide. As the mode propagates across the waveguide, it is strongly guided towards the area with a higher refractive index that leads to the mode shifting towards the outer rim of the bent waveguide. The maximum higher-order propagating mode of (2,2) was obtained with the higher intensity are found to be closer towards the inner wall of the waveguide. The propagating mode must fulfill this condition: $n_s < n_{eff} < n_w$ where n_{eff} denotes the effective refractive index. This value describes the propagation characteristic of the mode which proportional to propagation constant, β but reciprocal to wavenumber, κ_o. The maximum value

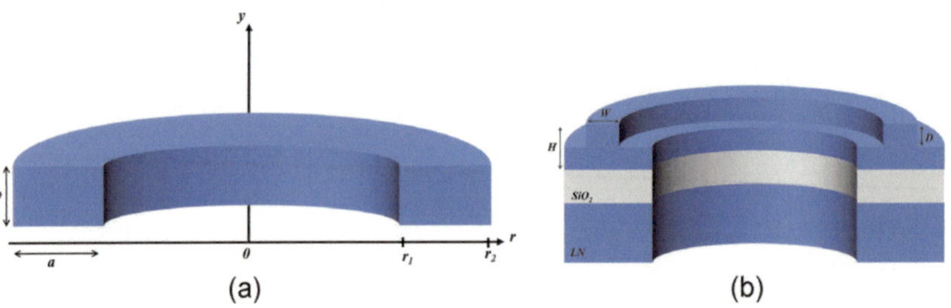

Figure 2.
(a) Toroidal waveguide with a width of core, $a = r_2 - r_1 = 1\ \mu m$ and height, $b = 0.5\ \mu m$ where $r_1 = 0.5\ \mu m$ and $r_2 = 1.5\ \mu m$. (b) LNOI rib waveguide with the thickness of the slab, $h = H - D$ [17].

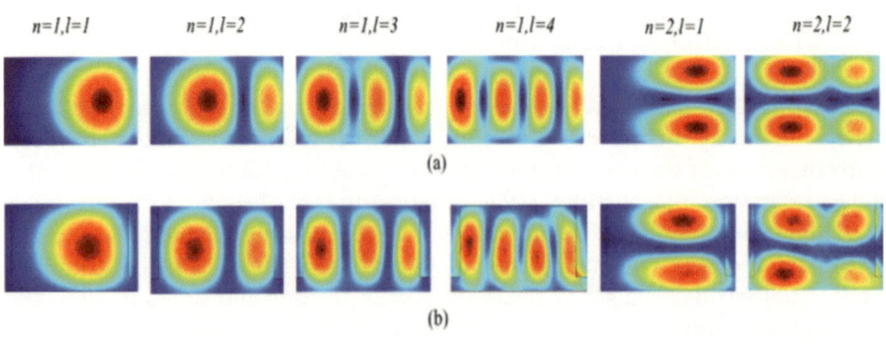

Figure 3.
Mode profile of bent waveguide in (a) toroidal and (b) LNOI rib structure [17].

of n_{eff} contribute to the eigenmode (fundamental) solution of Maxwell's equation. However, if the condition is not met ($n_{eff} < n_s$), the confined mode is degenerated and produced the radiating mode which contributes to the increment of propagation losses. Fortunately, the mode profile of the LNOI rib waveguide shows a prominent similarity with the toroidal waveguide structure despite the radius variation as depicted in **Figure 3(b)**. This output shows a strong consistency with the analytical solution in [17]. However, the slight asymmetric of the mode profile is led by the natural consequences of the symmetry breaking due to the presence of the oxide layer.

The aim of this chapter is to extend the above analysis by considering the variation of a radius on the rib waveguide prior to observe the deformation pattern of the mode profile within the bending area. Thus, the bending radius is set from 5 to 90 μm at a fixed width of the core value of 1 μm. From **Table 1**, it shows that the average radial position reduced as we c, which is inline with the finding in [17]. Consequently, the number of higher order mode increased as we increase the bending radius. However, at higher bending radius, the mode profile shows the symmetrical pattern compared to lower bending radius which tend to produce the asymmetric profile towards the outer surface of the waveguide.

Based on **Table 2**, the effective refractive index value decreased which reflect the decrement of propagation constant value. Simultaneously, the bending loss is increase drastically as the bending radius increase. The value is contributed by the summation of radiation, absorption, and scattering losses. However, it is mainly contributed by the scattering losses which occur due to the spatial fluctuations of the refractive index or the sidewalls roughness of the waveguide and radiation loss due to the imperfect mode guiding across the waveguide (**Table 2**) [18].

Mode number	Average radial position, r_{av}> (μm)											
	R = 5 μm		R = 10 μm		R = 30 μm		R = 50 μm		R = 70 μm		R = 90 μm	
	$l = 1$	$l = 2$	$l = 1$	$l = 2$	$l = 1$	$l = 2$	$l = 1$	$l = 2$	$l = 1$	$l = 2$	$l = 1$	$l = 2$
$n = 1$	1.199	1.3237	1.133	1.126	1.051	1.047	1.031	1.022	1.023	1.019	1.018	1.016
$n = 2$	1.014	—	0.977	0.982	0.987	1.00	0.988	1.102	0.997	1.003	0.995	1.002
$n = 3$	1.016	—	0.972	—	0.981	—	0.997	—	0.991	—	0.998	—
$n = 4$	—	—	0.900	—	0.995	—	0.996	—	0.980	—	0.900	—

Table 1.
The average radial position of LNOI rib waveguide for various bent radius.

Mode number	Effective refractive index, n_{eff}											
	R = 5 μm		R = 10 μm		R = 30 μm		R = 50 μm		R = 70 μm		R = 90 μm	
	$l = 1$	$l = 2$	$l = 1$	$l = 2$	$l = 1$	$l = 2$	$l = 1$	$l = 2$	$l = 1$	$l = 2$	$l = 1$	$l = 2$
$n = 1$	2.452	2.1574	2.331	2.125	2.223	2.022	2.207	2.012	2.201	2.000	2.197	1.997
$n = 2$	2.271	—	2.2087	2.0014	2.123	1.8542	2.096	1.842	2.090	1.837	2.087	1.834
$n = 3$	2.101	—	2.0603	2.0603	1.962	—	1.950	—	1.945	—	1.942	—
$n = 4$	—	—		1.8357	1.748	—	1.737	—	1.732	—	1.729	—

Table 2.
The effective refractive index of LNOI rib waveguide for various bent radius.

5. Conclusion

In summary, we have developed a numerical model solution based on the Helmholtz wave equation by using finite element method. The mode profiles of the numerical solution provide the information of optical signals in optical fiber. Thus we are shown the effect Quantum Anti-centrifugal force has a similar phenomenon as in the Helmholtz wave equation analytically. Note that, this is a consequence of the form of the Helmholtz equation which in this case is mathematically identical to the Schrödinger equation. Thus, the setups proposed in this work may provide a classical platform to test quantum phenomena in optical fiber. The development a numerical model of the bent structure for both toroidal and LNOI rib waveguide will serve as the best platform to conduct the fundamental investigation of this quantum fictitious force mainly in the classical platform. It allows one to predict the spatial mode number and profile accurately. From this numerical analysis, we can conclude that the eigenmode shows a fascinating behavior with their distortion which correlated with the geometry of the curvature.

Acknowledgements

The author acknowledges the financial support from Ministry of Higher Education and Universiti Malaysia Perlis. We also would like to extend our gratitude to Andrzej Gsjewski and Daniel Gustaw for their full technical support in realizing this numerical work.

Author details

Nor Roshidah Yusof, Norshamsuri Ali*, Syed Alwee Aljunid Syed Junid, Mohd Rashidi Che Beson and Rosdisham Endut
Faculty of Electronic Engineering Technology, Universiti Malaysia Perlis (UniMAP), Perlis, Malaysia

*Address all correspondence to: norshamsuri@unimap.edu.my

IntechOpen

© 2020 The Author(s). Licensee IntechOpen. This chapter is distributed under the terms of the Creative Commons Attribution License (http://creativecommons.org/licenses/by/3.0), which permits unrestricted use, distribution, and reproduction in any medium, provided the original work is properly cited.

References

[1] Xavier GB, Lima G. Quantum information processing with space-division multiplexing optical fibres. Communications on Physics. 2020;**3**:9. DOI: 10.1038/s42005-019-0269-7

[2] Agrawal, G. P. (2012). Fiber-optic communication systems (Vol. 222). John Wiley & Sons.

[3] Amnon Yariv, Pochi Yeh (2007). Photonics: Optical Electronics in Modern Communication 6th edition.

[4] Marcuse D. Field deformation and loss caused by curvature of optical fibers. Journal of the Optical Society of America. 1976;**66**:311-320. DOI: 10.1364/JOSA.66.000311

[5] Snitzer E. Cylindrical dielectric waveguide modes*. Journal of the Optical Society of America. 1961;**51**: 491-498. DOI: 10.1364/JOSA.51.000491

[6] Hu J, Menyuk CR. Understanding leaky modes: Slab waveguide revisited. Advances in Optics and Photonics. 2009;**1**:58-106. DOI: 10.1364/AOP.1.000058

[7] P. M. Morse and H. Feshbach, Methods of Theoretical Physics (McGraw-Hill Science/Engineering/Math, 1953).

[8] Marin L, Lesnic D, Mantič V. Treatment of singularities in Helmholtz-type equations using the boundary element method. Journal of Sound and Vibration. 2004;**278**(1–2):39-62. DOI: 10.1016/j.jsv.2003.09.059

[9] Marin L. Treatment of singularities in the method of fundamental solutions for two-dimensional Helmholtz-type equations. Applied Mathematical Modelling. 2010;**34**(6):1615-1633. DOI: 10.1016/j.apm.2009.09.009

[10] Cirone, M. A., Rzążewski, K., Schleich, W. P., Straub, F., & Wheeler, J. A. (2001). Quantum anticentrifugal force. Physical Review A, 65(2), 022101. DOI: 10.1103/PhysRevA.65.022101

[11] Białynicki-Birula I, Cirone MA, Dahl JP, Seligman TH, Straub F, Schleich WP. Quantum fictitious forces. Fortschritte der Physik. 2002;**50**(5–7): 599-607. DOI: 10.1002/1521-3978(200205)50:5/7<599::AID-PROP599>3.0.CO;2-G

[12] Dandoloff, R., Saxena, A., & Jensen, B. (2010). Geometry-induced potential on a two-dimensional section of a wormhole: Catenoid. Physical Review A, 81(1), 014102. DOI: 10.1103/PhysRevA.81.014102

[13] Dandoloff R, Atanasov V. Quantum anticentrifugal potential in a bent waveguide. Annalen der Physik. 2011; **523**(11):925-930. DOI: 10.1002/andp.201100136

[14] Atanasov, V., & Dandoloff, R. (2015). The curvature of the rotating disk and its quantum manifestation. Physica Scripta, 90, 065001. DOI: 10.1088/0031-8949/90/6/065001

[15] Dandoloff R. Topologically stable states of the anti-centrifugal potential. Journal of Modern Physics. 2019;**10**(08): 1002

[16] M. M. Milosevic, P. S. Matavulj, B. D. Timotijevic, G. T. Reed and G. Z. Mashanovich, "Design Rules for Single-Mode and Polarization-Independent Silicon-on-Insulator Rib Waveguides Using Stress Engineering, " in Journal of Lightwave Technology, vol. 26, no. 13, pp. 1840–1846, July1, 2008, DOI: 10.1109/JLT.2008.922193.

[17] Gajewski A, Gustaw D, Yusof NR, Ali N, Słowik K, Kolenderski P. Waveguide platform for quantum

anticentrifugal force. Optics Letters. 2020;**45**:3373-3376. DOI: 10.1364/OL.392216

[18] Yusof, N. R., Ali, N., Kolenderski, P., Hambali, N. A., & Aljunid, S. A. (2020, Jan). Geometrical optimization of lithium niobate on insulator rib waveguide for quantum communication application. In AIP Conference Proceedings (Vol. 2203, No. 1, p. 020064). AIP Publishing LLC. DOI: 10.1063/1.5142156

Chapter 3

Fabrication and Application of Polymer Optical Fiber Grating Devices

Rui Min

Abstract

Grating devices in polymer optical fiber (POFs) have attracted interest due to varies potential applications in recent years. This chapter presents the state of art about the fabrication technology of grating devices in different kinds of POFs and explores potential sensing application scenarios, focus on the fabrication of chirped POF FBG devices and the potential application of such devices. Present several typical applications with uniform POF FBG. Also present several typical applications based on Chirped POF FBG, which indicate POF FBG shown promising in the sensing area with show higher sensitivity and bio-compatibility than silica ones, and special grating in POF are attractive for future biomedical applications.

Keywords: polymer optical fiber, fiber Bragg grating, refractive index change, long-period grating

1. Introduction

Polymer Optical fiber (POF) is one kind of optical fiber made of polymer material. It has attractive characteristics such as low Young's modulus, high strain, high flexibility, and biocompatibility performance compared with silica optical fiber. The first commercially polymer optical fiber with deuterated polymer was reported in the early 1960s [1]. From the 1950s, optical fibers are under research due to the demand of transmission high-speed data. Silica fibers were also under intense research since their first proposal as a transmission medium by Kao et al. [2]. At that time the transmission loss of silica fibers was high and the theoretical predictions allow to propose pure glass as fiber material able to transmit light over a hundred kilometers. Then, the purity of silica material was significantly improved and losses were reduced up to 0.5 dB/km at 1300 nm, and 0.2 dB/km at 1550 nm after a few years [3]. Therefore, most of the research was shift to the silica optical fiber based transmission system, which forms the backbone of the modern telecommunications systems [4], the transmission loss of silica optical fiber, PMMA POF and CYTOP POF from 400 nm to 1600 nm wavelength shown in **Figure 1**.

The absorption loss of silica material is lower compared with polymer material although Rayleigh scattering is the main mechanism for both cases. The main absorption of polymer material contribution is due to the harmonics of the C-H vibration [5]. Researchers and companies in Japan did intense work to reduce the transmission loss of PMMA POF, also investigated perfluorinated polymer fibers (GI-POF) with

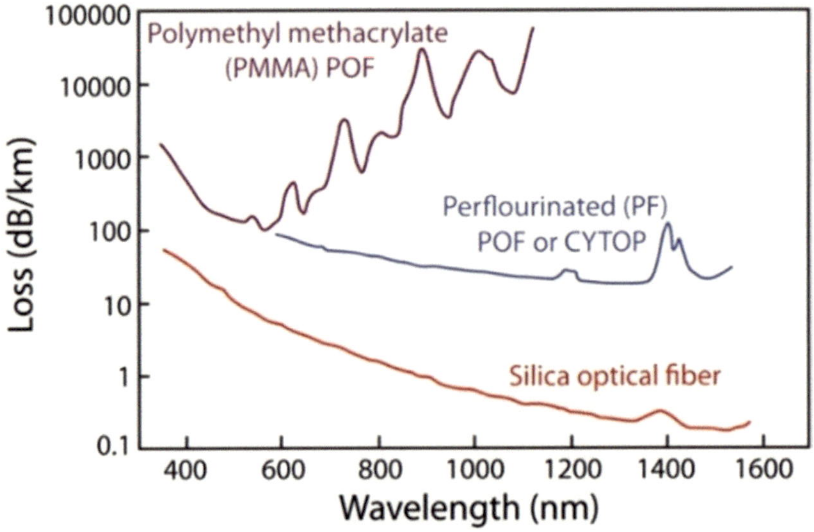

Figure 1.
The transmission loss of silica optical fiber, PMMA POF and CYTOP POF, and from 400 nm to 1600 nm wavelength.

lower losses, which shown 50 dB/km loss over 650 nm ~ 1300 nm wavelength [6–8]. However, the fabrication expend of the fluorinated polymer optical fiber are still higher compared with silica optical fiber. So, the most common material for polymer optical fiber is still PMMA which is cheap with high transmission loss at the telecom band. Both GI-POF and step-index POF mentioned above are multimode fibers, until now the main commercial POFs with single-mode performance with low fabrication cost still under research.

In recent years, the use of POFs for varies sensing applications under research are growing [9–13]. Sensing technology based on POF are mainly divided into intensity modulation technology, Brillouin scattering technology, Grating devices technology. Intensity modulation technology is the basic optical fiber sensing technologies, the main disadvantage is the performance related to the stability of the photo detector and the source, which always introduces errors in the measurement [14]. Brillouin scattering is an attractive sensing technology based on optical fiber which has been explored with silica optical fiber for many years [15]. POF with the advantage such as high flexibility can be exploited under large strains. Brillouin scattering technology in POF are good candidate for structural health monitoring. Grating devices technology as one of the main sensing technologies in silica fiber with high resolution and repeatable performance already goes from lab to the industry [16–18]. Although the first FBG in POF was reported by Peng's group with a step-index polymer optical fiber before twenty years [19], one main challenge for grating device fabrication in POF is the multimode performance of fiber, make it is difficult to analyses the spectrum from different modes.

The mature of photonic crystal fiber (PCF) technology and the excellent performance of PCF attracted the attention of the researcher for POF fabrication and application, which was invested with silica structures in 1997 [20]. The Large et al. obtained the first PCF with polymer material with endless single-mode performance in 2001 [21]. Then, H. Dobb et al. [22] reported the first FBG in endless single-mode mPOF in 2005, open the research of grating devices fabrication research with POF. Different materials such as Zeonex [23, 24], TOPAS [25, 26], and polycarbonate [27, 28] were used to fabricate microstructured polymer optical fiber (mPOF), different kind of laser irradiation technology were investigated for POG grating devices fabrication. Special grating such as Uniform FBG, Phase-shift FBG, Tilted FBG, Chirped FBG, LPG is investigated for various applications.

In this chapter, we report the basic theory of grating devices, and the lasted results about POF grating devices fabrication and potential applications. Section 2 will focus on the basic theory of grating devices such as FBG and LPG. Recent results on POF gratings fabrication with different kind of fibers and UV lasers will be reported in Section 3 and Section 4 will focus on introduce the potential applications of POF grating devices. Finally, the last section summarizes the main conclusions.

2. The basic theory of POF grating devices

Grating devices in optical fibers normally belong to two main types, which depending on the mode-coupling mechanism, counter-directional coupling is the responsible mechanism in fiber Bragg gratings (FBGs) and co-directional coupling as main mechanism for long-period gratings (LPG).

FBG is a device due to periodic modulation of the refractive index along the optical fiber, normally obtained through the exposure of the optical fiber core with an intense optical interference pattern [29]. The periodic structure in an FBGs as a selective mirror, and the wavelength satisfies the Bragg condition expressed as:

$$\lambda_{\text{Bragg}} = 2n_{\text{eff}}\,\Lambda \tag{1}$$

where n_{eff} is the average effective refractive index along the fiber and Λ is the grating period. Then, **Figure 2** shown a component reflects one wavelength and transmits all others with a wavelength-specific dielectric mirror response.

The ability of the FBGs shown optical communication applications such as band stop filters in Raman amplifiers [30] dispersion compensators [31], and wavelength division multiplexers [32]. Also, FBG can be used for sensing purposes, such as humidity, strain and temperature measurements.

As **Figure 3** shown, LPG is based on a periodic modulation of the refractive index of the fiber core along of the direction of propagation of the light, and the core mode couples to co-propagating cladding modes, different attenuation bands are obtained when the resonance condition is satisfied the equation below:

$$\lambda_{\text{res}} = \Lambda * \Delta \text{n} \tag{2}$$

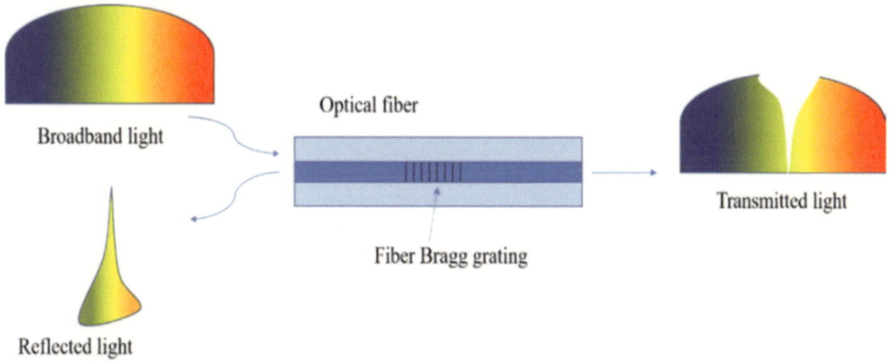

Figure 2.
The transmission and reflection spectrum of a broad light go through one fiber with FBG inside.

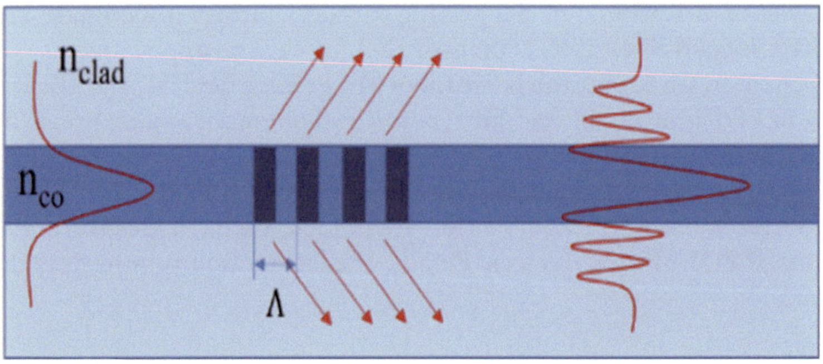

Figure 3.
LPG transmission graph.

Where $\Delta n = n_{co}$-n_c shown the difference between the effective refractive indices of the fundamental mode in core and the i th mode in cladding. As can be seen from the relation of the resonance wavelength, which is determined by the effective refractive indices of the core and cladding modes, so that any photo-induced, thermal-induced, geometrical, or mechanically induced periodic change will modify the position of the resonance wavelength. Which makes LPFGs are useful for the applications in fields as biological, chemical and optical sensing.

3. Fabrication of POF grating devices

POF FBG devices are normally obtained with direct writing technology [33], and phase mask technology [34]. Direct writing shown the advantage such as flexibility performance in terms of structure and wavelength, but the high cost femtosecond laser system is required for irradiation, and the resolution imposes limitations for the low wavelengths. LPG are normally obtained by heat imprinting [35], direct writing [36], and amplitude mask [37]. Here, we present several typical achievements obtained recently in the optimization of the fabrication process, grating devices inscription in a different kind of POF.

The fabrication process of POF grating devices has been reported by using different kind of lasers system such as 800 nm Ti:sapphire femto-second laser system [38], 532 nm Nd:YVO4 laser system [35], 387 nm Ti:sapphire femto-second laser system [39], 355 nm Nd:YAG laser system [40], 325 nm He-Cd laser [41], 266 nm Nd-YAG laser system [42] and 248 nm KrF excimer laser system [43]. Although 325 nm was the first irradiation wavelength for POF reported by the researchers in The University of New South Wales [44] and initially 248 nm and 266 nm wavelength were not considered suitable for POF grating writing due to high absorption of POF material at such wavelength, the first POF Bragg grating devices successful inscription less than one minutes with low flow and repetition rate 248 nm UV pulse opened a novel field of interest for POF grating irradiation [43]. Since then, the research about POF grating devices fabrication use 248 nm and 266 nm wavelength UV pulse has been continuously growing. A typical POF Bragg grating devices fabrication system is shown in **Figure 4**, where the pulse repetition rate and power can be optimized to modify the fabrication process, as shown in **Figure 4**.

Pure PMMA POF with low photosensitivity performance make it is difficlut to obtain strong POF grating device, doped POFs have attracted the researchers' interest to enhance photosensitivity performance. Researchers in The University of New South Wales employed a step-index multimode PMMA POF doped with organic dye in the core for grating devices irradiation with 325 nm wavelgnth UV

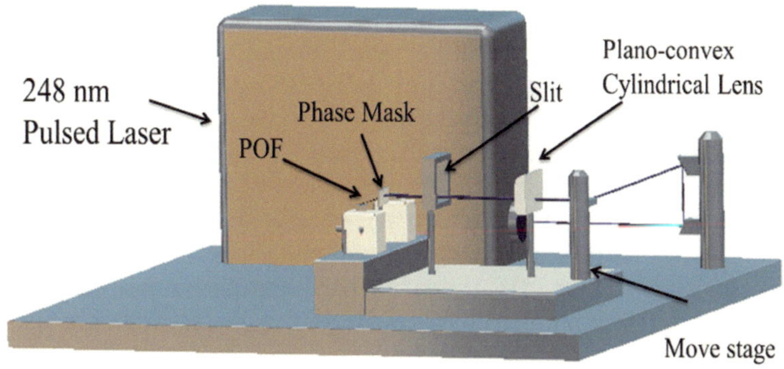

Figure 4.
Typical fabrication setup for POF FBG.

beam. The same group investigated benzyl methacrylate doped step-index multi-mode POF which lead a – 28-dB transmission spectrum with 85 mins exposure [45]. Researchers in The Hong Kong Polytechnic University investigated a step-index PMMA POF doped with TS (1% w.t.) and diphenyl sulfide (DPS) (5% mole) in the core and a pure 150 μm diameter PMMA cladding, obtained the FBG spectrum with −10 dB in transmission after 10 mins [46]. The researcher at Aston University reported a highly photosensitive mPOF doped with BDK in the core, FBG with −23 dB in the transmission was achieved after 13 mins [47]. The researchers in Belgium improved the optical fiber drawing technology with selected center hole doped with BDK in mPOF, a rapidly growing process of FBG with 83% reflectivity with 40 s [48]. Recently, the researcher in Hongkong investigated a new dopant material, named as diphenyl disulphide, which enables a fast and positive refractive index change with a low ultraviolet dose and leads to FBG fabrication with 7 ms under 325 nm wavelength UV beam irradiation [49].

The researchers optimized the fabrication process of FBGs in BDK doped POF with 248 nm and 266 nm wavelength, obtained strong grating with a single short pulse (15 and 8 ns of duration), the short time is even suitable for the fiber drawing process [42]. R. Min et al. also reported two, three, and five rings structure undoped PMMA POF fabricated with 266 nm Nd:YAG laser in the 850 nm region and using commercial ferrule connectors for POF connecting with silica fiber [50], the rend face and the reflected spectrum shown in **Figure 5**.

POF special grating devices such as tilted FBG and Chirped FBG are under research due to attractive applications. The first tilted POF FBG was investigated by the researcher's in Belgium, use of TS-doped photosensitive step-index PMMA POF with scanning phase mask technique, the transmitted amplitude spectrum evolution of a 3° angle is analyzed for the surrounding refractive index varies [51]. The first chirped POF FBG was irradiated with a KrF excimer laser operating at 248 nm wavelength and a 25-mm long chirped phase mask which customized for telecom-band FBG inscription in 2017 [52]. The laser pulse rate was 1 Hz and several shots were used for the grating response with a 1.2 nm/cm chirp and 3.9 nm bandwidth. The chirped phase mask technology offers good stability with the high cost and no flexibility as main drawbacks. Since then, different kind of techniques have been demonstrated for obtain chirped gratings in POF. The researchers in Cyprus used the femtosecond direct writing method for fabricate chirped FBG in commercial CYTOP POF [53], obtained 2000 periods with 10 nm bandwidth and a total length of around 4.5 mm. Femtosecond laser direct writing used for flexible chirped grating writing, although with the disadvantage as limit for low wavelengths. The first tunable chirped FBG was obtained with a tapered BDK doped mPOF by using a uniform phase mask under strain performance [54]. The spectral reflected power of a 10 mm bandwidth with a chirp of ~0.26 nm/mm under 1.6% strain, and the strain and temperature sensitivity obtained with 0.71 ± 0.02 pm/με and 56.7 pm/°C. Then, Chirped POFBGs have been also obtained by hot water-assisted gradient thermal annealing, where one grating device with around 1.1 nm/mm chirp performance was obtained as **Figure 6** shown [55]. The simplicity of this method is one of the main advantages since no special phase mask or additional etching is needed, and enables easy control of the chirp characteristics and the central wavelength.

Finally, regarding LPG in POF, the extensive literatures with different methods and mechanisms appeared in the last years. Recently, the researcher in Spain [56]

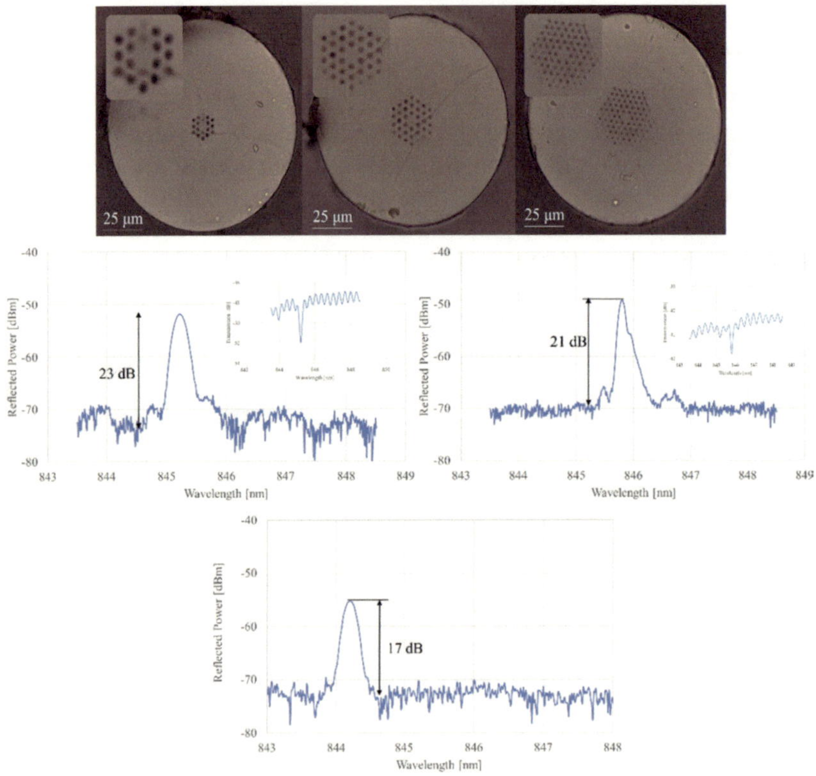

Figure 5.
The end face and the reflected spectrum of three types mPOF.

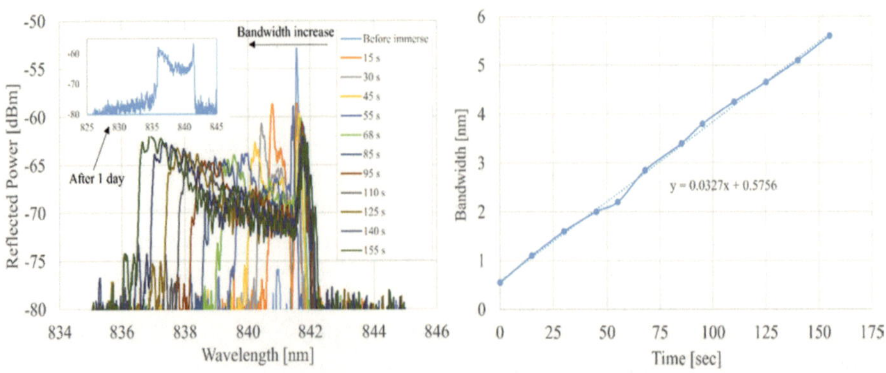

Figure 6.
The spectrum and the bandwidth varies of the chirped POF FBG fabrication with strong gradient annealing.

demonstrated a − 20-dB transmission LPG in mPOF with point by point method use a slit width of 0.2 mm; the beam was shifted 1 mm for inscribing each point and 25 steps were implanted. And the researchers in Cyprus reported an LPG in a CYTOP POF using a femtosecond laser inscription method. The LPG was inscribed directly in the center of the optical fiber core, tailored for operation at 1560 nm, which was characterized in transmission, and the response for relative humidity and temperature was also measured [57].

4. Applications of POF grating devices

POF FBG already goes to industry measurement such as used for water content detection in aviation fuel as shown in the **Figure 7**, the water content in Jet-A1 was measured by using POF FBG sensing technology which calibrated with both coulometric titration and environmental chamber. The results indicate a better performance compare with coulometric titration [58].

POF FBG can be also used to monitor the strain of human arteries with pulse wave signals. A variety of different vital signs including blood pressure can be derived from the signals, which show a higher signal to noise than silica FBG [59], the experimental measurement as **Figure 8** shown.

POF FBG can be used as health equipment for dynamic monitoring of gait. Five FBGs inscribed in CYTOP POF was embedded in a cork insole, as shown in **Figure 9**. The advantages of POF such as higher flexibility and robustness enabled monitoring patients with higher body mass, compared the results obtained with similar systems based on silica fiber, a mean sensitivity of ~8.14 PM/kPa was obtained, which is much higher compared with FBGs in silica optical fiber [60].

Consider the special POF grating applications, due to polymer special characteristics, strain sensing is the most attractive and reliable applications. There is a lot of literature reported POF FBG for strain sensing [61–63]. However, strain sensing under variable humidity and temperature conditions is always an issue for POF sensing technology go to real applications. The researchers in Spain demonstrated

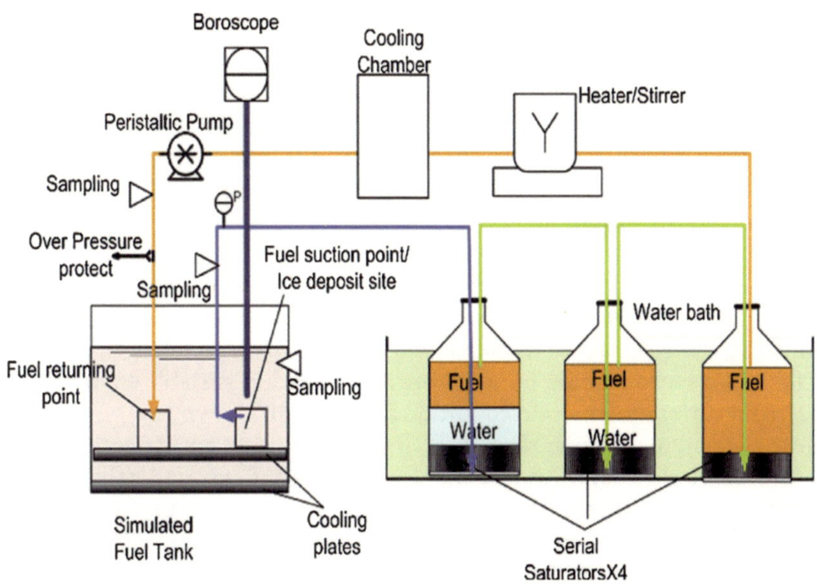

Figure 7.
Schematic of the bench test rig [58].

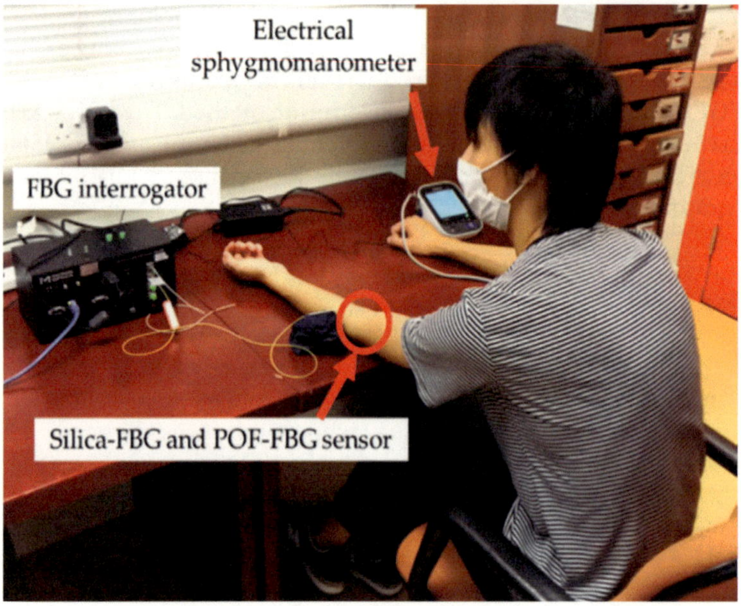

Figure 8.
Experimental measurement of blood pressure uses silica FBG and POF FBG [59].

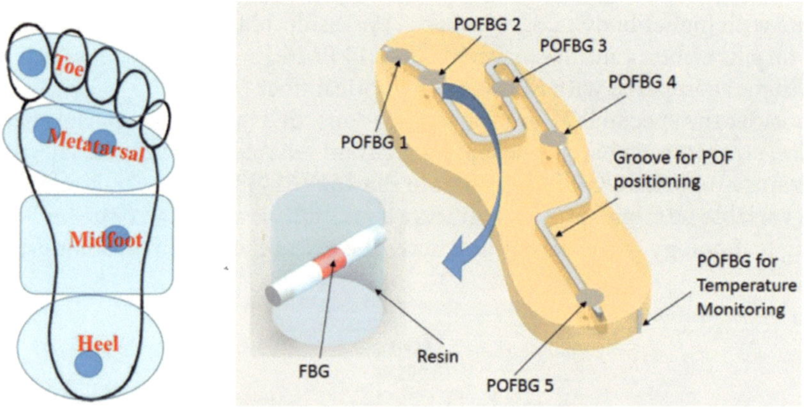

Figure 9.
Experimental measurement of blood pressure uses silica FBG and POF FBG. (a) Foot plantar area designation ans sensing point. (b) Polymer optical fiber Bragg grating embedded cork insole [60].

one method use the effective bandwidth of the tunable chirped POFBG, which is highly dependent on the strain and remains practically constant with temperature and humidity changes, can be implemented combine with wavelength measurement, for strain sensors under temperature and humidity variable environments, the spectrum varies under strain condition as **Figure 10** shown [64].

Due to the polymer characteristics, POF grating devices are attracting attention for biomedical applications. An essential feature of these systems is the possibility to detect temperature spatial distributions, which also name as thermal maps. A linearly chirped POF FBG reported as a semi-distributed temperature sensor for monitor the temperature profile along the grating length as **Figure 11** shown [65]. The grating device has been placed close to the radiofrequency applicator, which have one tip inserted in situ of the target. The reflection spectrum of the chirped

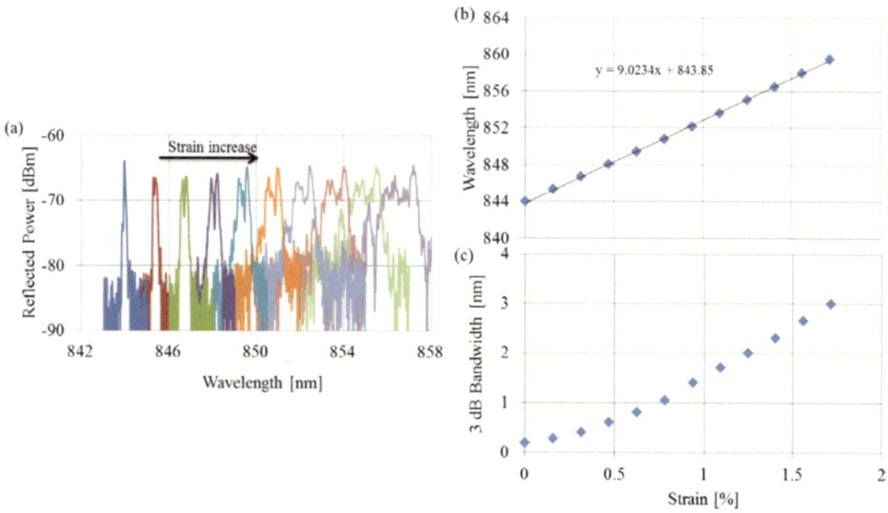

Figure 10.
Chirped POF FBG spectrum varies with strain. (a) Spectral reflected power vs strain; (b) Wavelength shift vs strain; (c) Bandwidth vs strain [64].

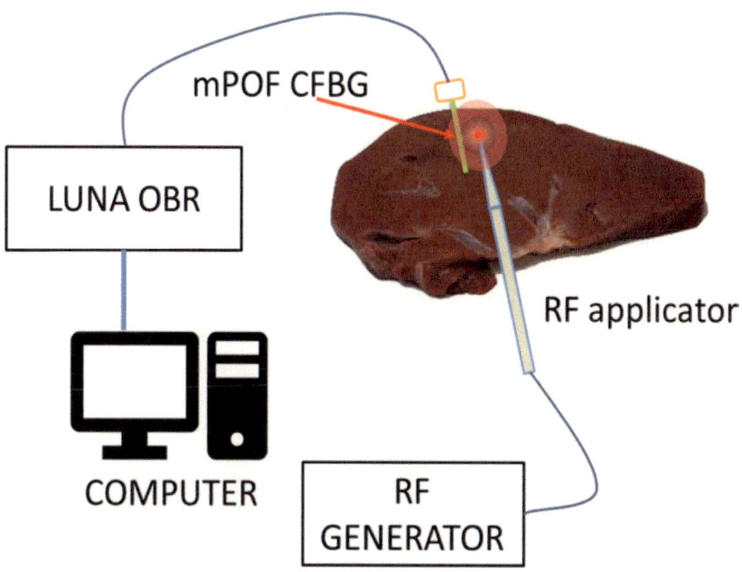

Figure 11.
Schematic of thermal ablation experiment [65].

POF FBG was detected by LUNA OBR 4600 and the temperature gradient was estimated use the Gaussian model method. The results indicate that chirped POF FBG can provide significant improvement in thermal detecting for biomedical applications.

Also, a high spatial resolution distributed strain sensing approach based on Chirped POF FBG was reported by researchers in China [66], through spatial wavelength encoded characteristic of Chirped POF FBG, a fully distributed high resolution strain measurement can be achieved by optical frequency domain reflectometry method, which is a promising approach for short-range fully distributed sensing systems, schematic of the experiment setup shown in **Figure 12**.

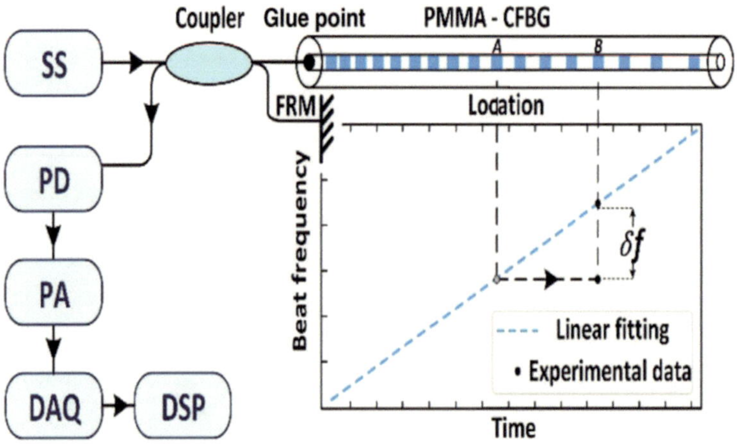

Figure 12.
Schematic of the experiment setup for fully distributed strain measurement [66].

5. Conclusion and outlook

Significant progress has been demonstrated for POF grating devices fabrication during the last years, such as to allow fast fabrication of POF grating devices under 248 nm and 266 nm wavelength UV, one 15 ns pulse for POF FBG fabrication. Besides the benefit as potential grating fabrication technology in the optical fiber drawing process, special grating structures such as Chirped FBG also take advantage of the short irradiation time, such as the benefit to reduce the stability requirements for the irradiation setup.

Besides strain and temperature sensitive devices as the main applications of POF FBG, special grating devices open new perspectives. As the main relevant examples, chirped POF FBGs used for high-resolution thermal detection in the biomedical area with a higher sensitivity and bio-compatibility than silica ones, special grating in POF are attractive for future biomedical applications, which also make special POF grating fabrication technology have room to improve.

To conclude, grating devices in POF show attractive performance for sensing applications. However, most of the POF for grating devices fabrication are still homemade, which need time and research to make this technology mature for real applications. From this point of view, grating devices fabrication in commercial CYTOP POF are promising, also the investigated for low cost single-mode commercial CYTOP POF and grating on such fiber.

Acknowledgements

The author thanks for the support of starting funding in Beijing Normal University at Zhuhai and the National Natural Science Foundation of China (62003046).

Conflict of interest

The authors declare no conflict of interest.

DOI: http://dx.doi.org/10.5772/intechopen.94351

Author details

Rui Min
Center for Cognition and Neuroergonomics, State Key Laboratory of Cognitive Neuroscience and Learning, Beijing Normal University at Zhuhai, Guangdong, China

*Address all correspondence to: ruimin@bnu.edu.cn

IntechOpen

© 2020 The Author(s). Licensee IntechOpen. This chapter is distributed under the terms of the Creative Commons Attribution License (http://creativecommons.org/licenses/by/3.0), which permits unrestricted use, distribution, and reproduction in any medium, provided the original work is properly cited.

References

[1] Bhowmik K, Gangding P. Polymer optical fibers. In: Peng GD, editor. Handbook of Optical Fibers. Springer. 2019. pp. 967-1017. DOI: 10.1007/978-981-10-1477-2

[2] Kao KC, Hockham GA. Dielectric-fibre surface waveguides for optical frequencies. Proc. Inst. Electr. Eng. 1966;**113**:1151-1158. DOI: 10.1049/piee.1966.0189

[3] Li MJ, Hayashi T. Advances in low-loss, large-area, and multicore fibers. In: Willner AE, editor. Optical Fiber Telecommunications VII. Academic Press. 2019. pp. 3-50. DOI: 10.1016/B978-0-12-816502-7.00001-4

[4] Shieh W, Djordjevic I. OFDM for Optical Communications. Academic Press. 2019:53-118. DOI: 10.1016/B978-0-12-374879-9.00003-4

[5] Groh, W. Overtone absorption in macromolecules for polymer optical fibers. Die Makromolekulare Chemie. 1988;**189**:2861-2874. DOI: 10.1002/macp.1988.021891213

[6] Ohtsuka Y, Hatanaka Y. Preparation of light-focusing plastic fiber by heat-drawing process. Applied Physics Letters. 1976;**29**:735-737. DOI: 10.1063/1.88921

[7] Koike Y, Kimoto Y, Ohtsuka Y. Studies on the light-focusing plastic rod. 12: The GRIN fiber lens of methyl methacrylate-vinyl Phenylacetate copolymer. Applied Optics. 1982;**21**:1057-1062. DOI: 10.1364/AO.21.001057

[8] Koike Y, Ishigure T. High-bandwidth plastic optical fiber for fiber to the display. J. Light. Technol. 2006;**24**:4541-4553. DOI: 10.1109/JLT.2006.885775

[9] Bilro L, Alberto N, Pinto JL, Nogueira R. Optical sensors based on plastic fibers. Sensors. 2012;**12**:12184-12207. DOI: 10.3390/s120912184

[10] Broadway C, Rui M, Leal Junior AG, Marques C, Caucheteur C. Towards commercial polymer fiber Bragg grating sensors: Review and applications. Journal of Lightwave Technology. 2018;**37**:2605-2615. DOI: 10.1109/JLT.2018.2885957

[11] Berghmans F, Geernaert T, Baghdasaryan T, Thienpont H. Challenges in the fabrication of fibre Bragg gratings in silica and polymer microstructured optical fibres. Laser and Photonics Reviews. 2014;**8**:27-52. DOI: 10.1002/lpor.201200103

[12] Min R, Ortega B, Marques C. Latest achievements in polymer optical fiber gratings: Fabrication and applications. MDPI Photonics. 2019; 6:36. DOI: 10.3390/photonics6020036

[13] Peters, K. Polymer optical fiber sensors—A review. Smart Materials and Structures, 2011; 20:013002. DOI: 10.1088/0964-1726/20/1/013002

[14] Lee B. Review of the present status of optical fiber sensors. Optical Fiber Technology. 2003;**9**:57-79. DOI: 10.1016/S1068-5200(02)00527-8

[15] Bao, X., Chen, L. Recent Progress in Brillouin scattering based fiber sensors. Sensors, 2011; 11: 4152-4187. DOI: 10.3390/s110404152

[16] Du, C., Dutta, S., Kurup, P., Yu, T., Wang, X. A review of railway infrastructure monitoring using fiber optic sensors. Sensors and Actuators A: Physical, 2020; 303: 111728. DOI: 10.1016/j.sna.2019.111728

[17] Qiao X, Shao Z, Bao W, Rong Q. Fiber Bragg grating sensors for the oil industry. Sensors. 2017;**17**:429. DOI: 10.3390/s17030429

[18] Chai Q, Luo Y, Ren J, Zhang J, Yang J, Yuan L, et al. Review on fiber-optic sensing in health monitoring of power grids. Optical Engineering. 2019;**58**:072007. DOI: 10.1117/1.OE.58.7.072007

[19] Peng GD, Xiong Z, Chu PL. Photosensitivity and gratings in dye-doped polymer optical fibers. Optical Fiber Technology. 1999;**5**:242-251. DOI: 10.1006/ofte.1998.0298

[20] Birks TA, Knight JC, Russell PSJ. Endlessly single-mode photonic crystal fiber. Optics Letters. 1997;**22**:961-963. DOI: 10.1364/OL.22.000961

[21] Eijkelenborg, Martijn A. van, Large, Maryanne C. J., Argyros, A., Zagari, J., Manos, S., A. Issa, N., Bassett, I., Fleming S., McPhedran, Ross C., Sterke, C. Martijn, and Nicorovici, Nicolae A.P. Microstructured polymer optical fibre. Opt. Express, 2001; 9: 319. DOI: 10.1364/OE.9.000319

[22] Dobb H, Webb DJ, Kalli K, Argyros A, Large MCJ, Eijkelenborg MA v. Continuous wave ultraviolet light-induced fiber Bragg gratings in few- and single-mode microstructured polymer optical fibers. Optics Letters. 2005;**30**:3296. DOI: 10.1364/OL.30.003296

[23] Woyessa G, Fasano A, Markos C, Stefani A, Rasmussen HK, Bang O. Zeonex microstructured polymer optical fiber: Fabrication friendly fibers for high temperature and humidity insensitive Bragg grating sensing. Optical Materials Express. 2017;**7**:286-295. DOI: 10.1364/OME.7.000286

[24] Woyessa G, Rasmussen HK, Bang O. Zeonex – A route towards low loss humidity insensitive single-mode step-index polymer optical fibre. Optical Fiber Technology. 2020;**57**:102231. DOI: 10.1016/j.yofte.2020.102231

[25] Markos C, Stefani A, Nielsen K, Rasmussen HK, Yuan W, Bang O. High-Tg TOPAS microstructured polymer optical fiber for fiber Bragg grating strain sensing at 110 degrees. Optics Express. 2013;**21**:4758-4765. DOI: 10.1364/OE.21.004758

[26] Yuan W, Khan L, Webb DJ, Kalli K, Rasmussen HK, Stefani A, et al. Humidity insensitive TOPAS polymer fiber Bragg grating sensor. Optics Express. 2011;**19**:19731-19739. DOI: 10.1364/OE.19.019731

[27] Fasano A, Woyessa G, Stajanca P, Markos C, Stefani A, Nielsen K, et al. Fabrication and characterization of polycarbonate microstructured polymer optical fibers for high-temperature-resistant fiber Bragg grating strain sensors. Optical Materials Express. 2016;**6**:649-659. DOI: 10.1364/OME.6.000649

[28] Zubel MG, Fasano A, Woyessa G, Min R, Leal-Junior A, Theodosiou A, et al. Bragg gratings inscribed in solid-core microstructured single-mode polymer optical fiber drawn from a 3D-printed polycarbonate preform. IEEE Sensors Journal. . DOI: 10.1109/JSEN.2020.3003469

[29] Hill KO, Meltz G. Fiber Bragg grating technology fundamentals and overview. Journal of Lightwave Technology. 1997;**15**. DOI: 10.1109/50.618320

[30] Sirleto L, Antonietta Ferrara M. Fiber amplifiers and fiber lasers based on stimulated Raman scattering: A review. Micromachines. 2020;**11**. DOI: 10.3390/mi11030247

[31] Petruzzi P, Lowry C, Sivanesan P. Dispersion compensation using only fiber Bragg gratings. IEEE Journal of Selected Topics in Quantum Electronics. 1999;**5**:1339-1344. DOI: 10.1109/2944.806759

[32] Dai Y, Liu Y, Leng J, Deng G, Asundi A. A novel time-division multiplexing fiber Bragg grating sensor interrogator for structural health monitoring. Optics and Lasers in Engineering. 2009;**47**:1028-1033. DOI: 0.1016/j.optlaseng.2009.05.012

[33] Lacraz A, Polis M, Theodosiou A, Koutsides C, Kalli K. Femtosecond laser inscribed Bragg gratings in low loss CYTOP polymer optical fiber. IEEE Photonics Technology Letters. 2015;**27**:693-696. DOI: 10.1109/LPT.2014.2386692

[34] Min R, Marques C, Bang O, Ortega B. Moiré phase-shifted fiber Bragg gratings in polymer optical fibers. Optical Fiber Technology. 2018;**41**:78-81. DOI: 10.1016/j.yofte.2018.01.003

[35] Xingsheng X, Hai M, Qijin Z. Properties of polarized laser-induced birefringent gratings in azobenzene-doped poly(methyl methecrylate) optical fibers. Optics Communications. 2002;**204**(1-6):137-143. DOI: 10.1016/S0030-4018(02)01198-7

[36] Kowal D, Statkiewicz-Barabach G. Microstructured polymer optical fiber for long period gratings fabrication using an ultraviolet laser beam. Optics Letters. 2014;**39**(8):2242-2245. DOI: 10.1364/OL.39.002242

[37] Hiscocks MP, van Eijkelenborg MA, Argyros A, Large MCJ. Stable imprinting of long-period gratings in microstructured polymer optical fibre. Optics Express. 2006;**14**(11):4644-4649. DOI: 10.1364/OE.14.004644

[38] Hu X, Kinet D, Chah K, Pun C-FJ, Tam H-Y, Caucheteur C. Bragg grating inscription in PMMA optical fibers using 400-nm femtosecond pulses. Optics Letters. 2017;**42**(14):2794. DOI: 10.1364/OL.42.002794

[39] Baum A, Scully PJ, Basanta M, Thomas CLP, Fielden PR, Goddard NJ, et al. Photochemistry of refractive index structures in poly(methyl methacrylate) by femtosecond laser irradiation. Optics Letters. 2007;**32**(2):190-192. DOI: 10.1364/OL.32.000190

[40] Liu HY, Peng GD, Chu PL, Koike Y, W Y. Photosensitivity in low-loss perfluoropolymer(CYTOP) fibre material. Electronics Letters. 2001;**37**(6):347-348. DOI: 10.1049/el

[41] Saez-Rodriguez D, Nielsen K, Bang O, Webb DJ. Photosensitivity mechanism of undoped poly (methyl methacrylate) under UV radiation at 325 nm and its spatial resolution limit. Optics Letters. 2014;**39**(12):3421-3424. DOI: 10.1364/OL.39.003421

[42] Pereira L, Min R, Hu X, Caucheteur C, Bang O, Ortega B, et al. Polymer optical fiber Bragg grating inscription with a single Nd:YAG laser pulse. Optics Express. 2018;**26**(14):18096-18104. DOI: 10.1364/OE.26.018096

[43] Oliveira R, Bilro L, Nogueira R. Bragg gratings in a few mode microstructured polymer optical fiber in less than 30 seconds. Optics Express. 2015;**23**(8):10181-10187. DOI: 10.1364/OE.23.010181

[44] Xiong Z, Peng GD, Wu B, Chu PL. Highly tunable Bragg gratings in single-mode polymer optical fibers. IEEE Photonics Technology Letters. 1999;**11**(3):352-354. DOI: 10.1109/68.748232

[45] Liu HY, Peng GD, Chu PL. Polymer fiber Bragg gratings with 28-dB transmission rejection. IEEE Photonics Technology Letters. 2002;**14**(7):935-937. DOI: 10.1109/LPT.2002.1012390

[46] Yu J, Tao X, Tam H. Trans-4-stilbenemethanol-doped photosensitive polymer fibers and gratings. Optics Letters. 2004;**29**(2):156-158. DOI: 10.1364/OL.29.000156

[47] Sáez-Rodríguez D, Nielsen K, Rasmussen HK, Bang O, Webb DJ. Highly photosensitive polymethyl methacrylate microstructured polymer optical fiber with doped core. Optics Letters. 2013;**38**(19):3769-3772. DOI: 10.1364/OL.38.003769

[48] Hu X, Woyessa G, Kinet D, Janting J, Nielsen K, Bang O, et al. BDK-doped core microstructured PMMA optical fiber for effective Bragg grating photo-inscription. Optics Letters. 2017;**42**(11):2209-2212. DOI: 10.1364/OL.42.002209

[49] Bonefacino J, Tam H-Y, Glen TS, Cheng X, Pun C-FJ, Wang J, et al. Ultra-fast polymer optical fibre Bragg grating inscription for medical devices. Light: Science & Applications. 2018;**7**(3):17161. DOI: 10.1038/lsa.2017.161

[50] Min R, Pereira L, Paixao T, Woyessa G, Andre P, Bang O, et al. Inscription of Bragg gratings in undoped PMMA mPOF with Nd:YAG laser at 266 nm wavelength. Optics Express. 2019;**27**(26):38039-38048. DOI: 10.1364/OE.27.038039

[51] Hu X, Pun C-FJ, Tam H-Y, Mégret P, Caucheteur C. Tilted Bragg gratings in step-index polymer optical fiber. Optics Letters. 2014;**39**(24):6835-6838. DOI: 10.1364/OE.22.018807

[52] Marques CAF, Antunes P, Mergo P, Webb DJ, Andre P. Chirped Bragg gratings in PMMA step-index polymer optical fiber. IEEE Photonics Technology Letters. 2017;**29**(6):500-503. DOI: 10.1109/LPT.2017.2662219

[53] Theodosiou A, Hu X, Caucheteur C, Kalli K. Bragg gratings and Fabry-Perot cavities in low-loss multimode CYTOP polymer fiber. IEEE Photonics Technology Letters. 2018b;**30**(9):857-860. DOI: 10.1109/LPT.2018.2820381

[54] Min R, Ortega B, Marques C. Fabrication of tunable chirped mPOF Bragg gratings using a uniform phase mask. Optics Express. 2018;**26**(4):4411-4420. DOI: 10.1364/OE.26.004411

[55] Min R, Ortega B, Broadway C, Caucheteur C, Woyessa G, Bang O, et al. Hot water-assisted fabrication of chirped polymer optical fiber Bragg gratings. Optics Express. 2018;**26**(26):34655-34664. DOI: 10.1364/OE.26.034655

[56] Min R, Marques C, Nielsen K, Bang O, Ortega B. Fast inscription of long period gratings in microstructured polymer optical fibers. IEEE Sensors Journal. 2018;**18**(5):1919-1923. DOI: 10.1109/JSEN.2018.2791663

[57] Theodosiou A, Min A, Junior A, Ioannou A, Neto A, Pontes M, et al. Long period grating in a multimode CYTOP polymer fibre inscribed using a femtosecond laser. Optics Letters. 2019;**44**(21):5346-5349. DOI: 10.1364/OL.44.005346

[58] Zhang W, Webb DJ, Lao L, Hammond D, Carpenter M, Williams C. Water content detection in aviation fuel by using PMMA based optical fiber grating. Sensors and Actuators B: Chemical. 2019;**282**(1):774-779. DOI: 10.1016/j.snb.2018.11.134

[59] Haseda Y, Bonefacino J, Tam H-Y, Chino S, Koyama S, Ishizawa H. Measurement of pulse wave signals and blood pressure by a plastic optical fiber FBG sensor. Sensors. 2019;**19**:5088. DOI: 10.3390/s19235088

[60] Vilarinho D, Theodosiou A, Leitão C, Leal-Junior AG, Domingues MDF, Kalli K, André P, Antunes P, Marques C. POFBG-embedded cork insole for plantar pressure monitoring. Sensors. 2017;**17**:2924

[61] Zhang ZF, Zhang C, Tao XM, Wang GF, Peng GD. Inscription of polymer optical fiber Bragg grating

at 962 nm and its potential in strain sensing. IEEE Photonics Technology Letters. 2010;**22**(21):1562-1564. DOI: 10.1109/LPT.2010.2069090

[62] Webb DJ. Fibre Bragg grating sensors in polymer optical fibres. Measurement Science and Technology. 2015;**26**(9):092004. DOI: 10.1088/0957-0233/26/9/092004

[63] Leal-Junior, A., Theodosiou, A., Frizera-Neto, A., Pontes, M. J., Shafir, E., Palchik, O., ... & André, P. (2018). Characterization of a new polymer optical fiber with enhanced sensing capabilities using a Bragg grating. Optics Letters, 43(19), 4799-4802. DOI:10.1364/OL.43.004799

[64] Min R, Ortega B, Broadway C, Hu X, Caucheteur C, Bang O, et al. Microstructured PMMA POF chirped Bragg gratings for strain sensing. Optical Fiber Technology. 2018;**45**:330-335. DOI: 10.1016/j.yofte.2018.08.016

[65] Korganbayev S, Min R, Jelbuldina M, Hu X, Caucheteur C, Bang O, et al. Thermal profile detection through high-sensitivity fiber optic chirped Bragg grating on microstructured PMMA fiber. Journal of Lightwave Technology. 2018;**36**(20):4723-4729. DOI: 10.1109/JLT.2018.2864113

[66] Lyu C, Liu Z, Huo Z, Ge C, Cheng X, Tam H-Y. High-sensitivity, high-spatial-resolution distributed strain sensing based on a poly (methyl methacrylate) chirped fiber Bragg grating. Photon. Res. 2020;**8**:1134-1139. DOI: 10.1364/PRJ.391160

Chapter 4

Modified Single Mode Optical Fiber Ammonia Sensors Deploying PANI Thin Films

Husam Abduldaem Mohammed and Mohd Hanif Yaacob

Abstract

Modified optical fiber sensors received increasing attention because of their superior properties over electrical sensors. These properties include their immunity towards electromagnetic interference and the ability to be deployed in corrosive and volatile environment. Several optical fiber platforms have been developed for chemical sensing applications based on modifying optical fiber cladding layer such as etched, tapered, D-shaped and etched-tapered. The modifications purpose is to extend the evanescent wave propagating out of the core physical dimensions. Thus, evanescent wave interaction with analyte is enhanced. Modified optical transducing platforms are integrated in gas sensing applications, such as ammonia. Modified optical fiber sensors coated with nanostructured thin films have been developed and gained popularity as practical devices towards gases with low concentrations. The development and characterization of the modified SMF sensing platforms including etched, tapered and etched-tapered platforms against ammonia will be presented in this chapter. These platforms were coated with PANI nanostructured thin film. The 50 μm etched-tapered SMF coated with PANI produced response, recovery times, and sensitivity of 58 s, 475 s, and 231.5%, respectively, in the C-band range. The limit of detection of the modified fiber sensor was 25 ppm. The developed sensors exhibit good repeatability, reversibility, and selectivity.

Keywords: optical fiber sensors, ammonia sensors, gas sensors, polyaniline, modified SMF sensors, etched-tapered optical fiber, C-band sensors

1. Introduction

The conventional ammonia (NH_3) sensors are electrical type. The electrical sensors have simple structure and low cost but they have poor selectivity as they respond to other gases. Moreover, electrical sensors are susceptible towards electromagnetic interference. Because of signal ignition opportunity, the electrical sensors are not appropriate to be used in oil and gas volatile environment [1]. There is a critical demand to develop an alternate type of sensors to avoid disasters resulted from ammonia leakages or drawbacks related to electrical signal based sensors {Mohammed, 2019 #3943}. The optical fiber sensor is an outstanding alternate [2]. Mostly, modified multimode optical fiber (MMF) is deployed to fabricate current NH_3 optical fiber sensors. Generally, the MMF sensors have lower sensitivity than the single optical fiber SMF sensors that not extensively explored for NH_3 sensing [2].

IntechOpen

Researchers showed intensive focus on modified optical fiber platforms as sensing tools since they are more sensitive compared to the conventional fibers. Cladding modified SMF sensors with high sensitivity integrated with nanostructured thin films against ammonia can be deployed to avoid crises resulted from gas leakage such as ammonia [2]. These sensors have been gained popularity as practical tool to detect chemicals with low concentrations such as gases. By utilizing these configurations, it is expected to fabricate sensors with high sensitivity and fast response.

The aim of this chapter is to design and demonstrate an etched-tapered SMF optical fiber gas sensor for remote monitoring application. The gas under testing is ammonia due to its high severity and deployment in the industry. The objectives to achieve this research project are as follows:

- To present different modified optical fiber transducing platforms and nanomaterials used as sensing layers particularly, polyaniline.

- To design, fabricate and characterize modified SMF transducing platforms, that are etched, tapered and etched-tapered SMF platforms.

- To synthesize, deposit, characterize and evaluate gas sensing characteristics of the PANI nanofiber as a sensing layer onto developed modified SMF transducing platforms towards different concentrations of NH3 gas within C-band wavelength ranges.

- To compare performance criteria of developed SMF sensors with the reported sensors in some previous studies.

The next section presents a description of the modified optical fiber platforms as sensing tools since they are more sensitive compared to the conventional fibers. The etched, tapered, etched-tapered platforms as modified optical fiber platforms will be elaborated. After that, the Polyaniline nanofiber employed as a sensing layer is introduced in details. The properties of PANI thin films will be discussed by highlighting its attraction and factors that influence the sensing performance. Later, a detailed review on previous works that use PANI as a sensing layer for ammonia sensors will be presented. Moreover, PANI nanostructured thin film preparation and deposition onto the SMF transducing platforms will be highlighted. Several micro-characterizations of the fabricated nanostructured thin films were carried out to investigate sensing layer morphology and thickness of the nanostructured thin film. These parameters affect gas sensing performance. The sensing performance of the modified SMF sensors including etched, tapered and etched-tapered sensors coated with sprayed PANI nanofibers will be investigated and analyzed when exposed to ammonia with different concentrations. The investigation performed in the range of C-band wavelengths at room temperature. Based on author's knowledge, the investigation of the SMF sensors coated with PANI in C-band wavelengths ranges is not explored yet. Finally, chapter conclusions will be summarizing the performance properties behind the deployment of modified SMF sensing platforms Integrated with PANI nanofibers at room temperature.

2. Modified single mode optical fiber (SMF) sensors

Researchers showed intensive focus on modified optical fiber platforms as sensing tools since they are more sensitive compared to the conventional fibers. The high sensitivity in the modified fiber is a consequence of the evanescent wave or some portion of the optical power propagates outside of the core layer [3].

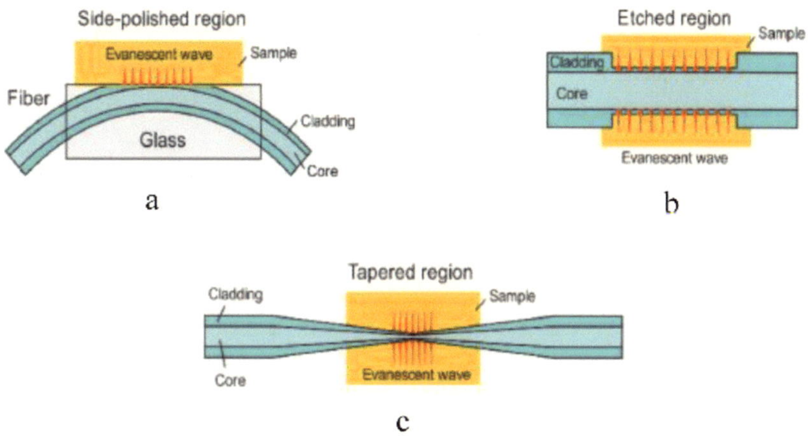

Figure 1.
Modified optical fiber sensors (a) side-polished, (b) etched and (c) tapered fibers [10, 11].

Evanescence wave-based intensity sensors require fiber modification to expand the penetration depth of the evanescence wave to increase its interaction with the surroundings. As a result, different optical fiber modification techniques are deployed including side polishing [4], D-fibers [5], chemical etching [6, 7] or tapering [8, 9] as shown the **Figure 1** [10] to improve the evanescence field for sensing applications [11].

2.1 Etched optical fiber sensor

Etched optical fiber is widely used for evanescent based sensors. The common process is to etch the cladding part of the fiber. One way to remove the cladding is to immerse the optical fiber using a strong acid in a chemical bath. Chemical etching produces shorter tapers with larger cone angles, resulting in higher transmission efficiency [12]. Depending on the composition of the glass, different acidic solutions are used. The etching rate depends on the dopant concentration in the structure of the optical fiber and concentration of the chemical solution. Etching parameters such as solvent type, acid concentration, etching time, and temperature are critical factors for the resultant optical and geometrical characteristics of the optical fiber [11].

Even though the modified optical fiber sensors enhance the evanescent wave to interact with the surrounding, each technique has its drawbacks. Side-polished fiber is made through polishing a fiber that is implanted on a block, such as a quartz block. The weakness of this technique is that the long time consumed in the fabrication procedures and it is difficult to produce a long sensing region. Hence, it difficult to develop a high sensitivity sensor [13]. The D-shape fiber is an optical fiber with removing half of the cladding layer. It has an advantage of long evanescent field interaction length. The removed-clad, namely etched fiber, sensor offers a simple and inexpensive fabrication method, especially compared with mechanical pulling and D fibers. On the other hand, chemical etching process is not easily controllable. The fabrication of tapered optical fibers is more reproducible and the controllable with the advance in tapering machine technology. However, the tapering machine itself is very expensive.

2.2 Tapered optical fiber sensors

In tapered optical fiber, the fiber diameter is reduced at a specific area called taper waist through heating and pulling the fiber ends in opposite directions as

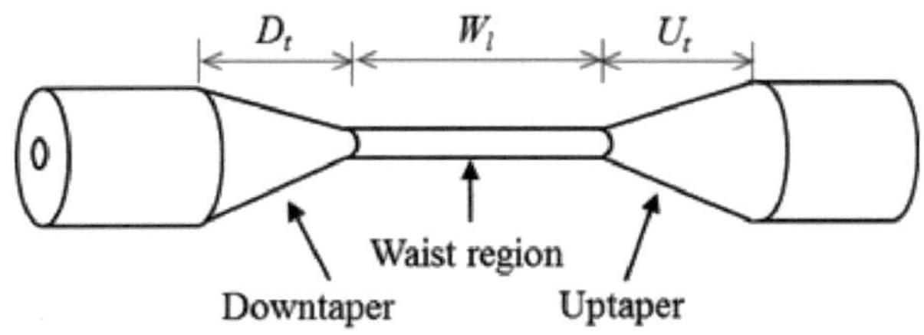

Figure 2.
Demonstration of tapered optical fiber [11].

shown in **Figure 2** [11, 14]. Based on the figure, D_t, W_l and U_t represents the down taper, waist length and up taper regions which are the tapered fiber profile [11].

Several techniques have been developed to fabricate the tapered optical fibers such as etching [15] and flame heating. The flame heating technique has been verified to be the most flexible technique that results in robust physical properties. Later, the flame was replaced with microheater that is more controllable. Recently, computer-controlled machines that are able to fabricate tapers with desired dimensions are available in the market. Conventional SMF or multimode optical fiber (MMF) with standard diameter of 125 μm can be tapered down to 5 μm using the machine [16, 17].

Tapering process improves the sensitivity of the optical fiber sensors by easing the access to the evanescent field, which enables strong interaction between the light and the analyte. To prepare a qualified tapered fiber based devices, the tapered fibers used should be fabricated with high adiabaticity, uniform microfiber diameter and suitable microfiber diameter with large evanescent wave. Fundamentally, strong evanescent wave is obtainable with thinner tapering fiber diameters. Accordingly, the tapered fibers are made with small diameter in the range of 0.8–3 μm for most devices uses tapered fibers.

The strong evanescent wave on the taper waist surface make it more sensitive to its surrounding. The optical fiber sensors fabricated either by etching or tapering processes can be coated with suitable nanostructured sensing layer in order to improve the sensitivity. When sensing layer reacts towards the target analyte, its optical properties may change. Hence, the amount of evanescent wave absorbed by the sensing layer is changed according to the analyte concentrations [11].

2.3 Etched-tapered optical fiber sensors

The Etched-tapered optical fiber platform comprises of the etching and tapering processes abovementioned. Firstly, hydrofluoric acid is used to remove some of the cladding of the SMF as illustrated previously. The etched area is then tapered using different tapering methods based on the recommended configurations. Dealing with etched fibers to perform the tapering process is a critical and more challenging as compared to tapering the standard fiber [11]. The removal of the cladding layer of the SMF increases its fragility and increase difficulty of fiber handling. This is may be overcome by utilizing a customized holder in which the etched fiber is positioned while it is tapered.

Optical fiber sensors are deployed for detection of different hazardous gases including ammonia (NH_3). Ammonia is widely used gas in different applications such as chemical industries, agricultures and medicines [18, 19] . Natural NH_3 level present in the atmosphere is in low ppb (1–5 ppb) levels. NH_3 can be characterized by its colorless, pungent smell, and explosive, toxic at a high-concentration NH_3 atmosphere [20, 21]. Generally, upon exposure to around 50 ppm NH_3 gas in air may cause acute poisoning or life-threatening situations such as permanent blindness, lung disease, respiratory disease, skin disease. The Occupational Safety and Health Administration (OSHA) has set a limit of 25 ppm in the workplace throughout an 8 hours shift and a short-term limit (15 min) of 35 ppm [11, 22]. Additionally, NH_3 sensors remain the potential candidates employed in agriculture, chemical industries, pharmaceutical, hydrogen fuels, defense and food processing industries to monitor the NH_3 leak in controlled atmospheres. Hence, the development of highly-sensitive and reliable NH_3 sensors to continuously detect leakages NH_3 is a key issue for the safety of environments [19].

More worrying, the gas is flammable at 50°C at very high concentration (150,000 ppm) [11, 23]. In chemical leakage alarm for NH_3 detection, the detection limit ≥1000 ppm with operating temperature range up to 500°C and required response time may be in the range of minutes. The concentrations can be very high at NH_3 plants and can even be explosive [24]. Recently, on Aug 2016, Petronas Chemicals Group stated that two workers were killed and three injured by an ammonia leak at a Malaysian chemical plant [25]. Hence, the development of highly-sensitive and reliable sensors to continuously detect leakages of NH_3 is a key issue for the safety of environments [19, 26–29].

In general, the performance of the nanomaterial based sensors, particularly sensitivity, is controlled by three factors, material characteristics, transducer function and variability for sensor development [30]. The material characteristics indicate its surface ability to detect a specified chemical. The transducer function refers to the ability of converting the response of interaction between the analyte and nanomaterial surface into readable signal [31]. For the purpose of effectively upgrading the performance of the sensor, the surface properties might be enhanced by depositing nanostructured thin film as a sensing layer. In nanometer dimensions, the majority of the particles (atoms) are surface or near surface of the sensing platform. Therefore, the effective number of existing sites to interact with analyte molecules is high [11]. The deployment of nanomaterial sensing layers reduces the size of the detecting parts and transducer as well as reduced cost and response time. This results in scaling down of the detecting devices and simplicity. Moreover, nanomaterial sensing layers provides high surface to volume ratio leads to better detection limits [32]. The sensors incorporating the nanostructured sensing layer including conducting polymers such as polyaniline (PANI) has showed ability to integrate with different transducing platforms [11, 32].

3. Polyaniline nanostructures

Conducting polymers have become popular since early 1980s [33] due to their low cost, ease of synthesis and processing with ability to sense in room temperature [34]. Polyaniline (PANI) is of the important conducting polymers exploited extensively and studied as sensing materials. The light weight, high conductivity, mechanical flexibility and low-cost leads to the use of PANI in many applications.

PANI exist in several oxidation states with different colors. Generally, the fundamental form of polyaniline known as emeraldine. Emeraldine forms can either be in emeraldine base (EB) or protonated emeraldine salt (ES) forms. Reducing emeraldine base generates the leucoemeraldine base (LEB) or pernigraniline base (PEB) in oxidized forms. [35]. The acid/base doping response makes PANI attractive for acid/base chemical vapor sensors, super capacitors, as well as biosensors. Potential aspects of PANI make it promising for sensing applications since it is presents different oxidation states each with different color, changes and conformations [11].

PANI-ES is the only conducting form of PANI with approximately 15 S cm^{-1} conductivity. Meanwhile, other forms are normally insulating with conductivity below 10–5 S cm-1 [36, 37]. The PANI EB and ES form can be identified through their colors, where EB is blue and ES is green [36]. PANI-ES can be obtained through doping process, either by oxidation of leucoemeraldine base or by protonation of the PANI-EB [36]. The protonation is carried out by processing the PANI-EB with a strong acid such as HCl that induces the protonation of the imine sites.

PANI is attractive to be used as a sensing layer because it can rapidly switch between the EB and ES forms as it is exposed to certain analytes. This reversible process is also known as doping (ES) or dedoping (EB). This reversible pH-switching property not only changes its electrical conductivity, but also its optical property. The change in optical properties can be observed through the change in the absorbance spectrum.

PANI has been proposed for sensing NH_3 since there were variations in the electrical conductivity and optical absorption on exposure to NH_3. The properties change with the condition of oxidation and protonation of the polymer. At the point when exposing PANI-ES (the acid form) to NH_3, it will be deprotonated and transferred into a non-conducting PANI-EB [11, 21]. While there are many reported studies on the PANI based electrical sensors [38–41], the optical fiber based NH_3 sensors employing PANI is not as popular as the electrical one [42, 43].

Sensors that use PANI in nanostructure forms such as nanofibers or nanorods have shown a significantly better performance in terms of response time and sensitivity compared to the ones that use conventional PANI films [44]. This is as a result of increased surface area, high porosity, and small structure diameter which enhances the diffusion of the analyte molecules into the nanostructures [44]. PANI nanofibers can be obtained through various methods such as template synthesis, phase separation and electrospinning [45]. Several approaches have been adopted to enhance the PANI sensing performance (sensitivity and selectivity) [46]. This includes polymer molecular structures modification, using different dopants, and integrating the conducting PANI with different types of inorganic materials such as graphene-like materials [47]. The conductivity of PANI can be also enhanced using a highly conductive filler as graphene and graphite [48].

4. Review of ammonia sensors based on polyaniline

Limited SMF based NH_3 sensors employing PANI nanocomposite have been proposed so far. The developed optical sensors utilized a few types of substrates including glass substrate, waveguide, and modified optical fibers. Different optical measurement techniques such as absorption, transmittance, reflectance, resonance wavelength shift and fluorescence are used in the development of NH_3 optical sensors coated with PANI. The development of NH_3 optical sensors coated with PANI can be carried out using different deposition methods. This includes in-situ deposition, drop casting, dip coating, spray, electrochemical deposition, or spin coating.

The influence of synthesis methods, deposition methods, dopant types on PANI morphology and NH_3 sensing properties was studied in [49]. Glass substrate was used and absorbance measurement was done at wavelength of 632 nm. They experimented with three synthesis methods (interfacial, rapid mixing and dropwise mixing), two deposition methods (in-situ and drop-coating) with three types of dopants (HCl, CSA, and I_2). The results demonstrated that in-situ deposited PANI formed a cauliflower-like nanoparticles structure with a thickness of approximately 400 nm and diameter of 300 nm. While in the case of drop-casted PANI, a PANI nanofibers was formed with measured diameters of approximately

60–90 nm and 350 nm length. The in-situ deposited PANI nanostructure showed a shorter response time with higher sensitivity compared to drop-casted PANI, mainly because of more uniform coating. Drop-casting method suffered from the problem of non-uniform coating due to agglomeration of the PANI at certain areas on the glass surface. In-situ deposition of PANI-HCl was used for other experiments to investigate the best synthesis method. Rapid mixing method with oxidant-to-monomer mole ratio of equal to 1 was found to give the best result. This was contributed from the highest porosity and highest surface area of PANI nanogranules with size of 200 nm – 300 nm. PANI-CSA was found to give the highest sensitivity and the most stable NH_3 sensor as compared to PANI-HCl and I_2-doped PANI. The UV–Vis and FTIR measurements also confirmed that the sensing mechanism is based on the deprotonation process. From this work, it is learnt that to achieve high sensitivity, the highest surface area nanostructure is desirable together with high doping level.

In [50], a super Fiber Bragg grating (FBG) NH_3 sensor was developed based on optical reflectance measurement method. The FBG sensor was fabricated by removing some part of its cladding using chemical etching with 14 μm reduced diameter. The PANI sensing layer was deposited on the etched area with 300 nm thickness. The FBG showed a blue shift in Bragg's wavelength towards the shorter wavelengths as exposed to higher NH_3 concentration with 0.073 pm/ppm sensitivity. The value is low so that it is hardly to be measured.

Surface plasmon resonance-based NH_3 plastic optical fiber sensor coated with PANI as sensing layer was reported in [51]. The sensor was fabricated by uncladding 1 cm length of a 600 μm fiber diameter. The unclad area was coated with different thickness of indium tin oxide (ITO) and PANI on top with the use of thermal evaporation technique. Then, the ITO coated fiber was dipped into ammonium hydroxide (as adhesive), followed by PANI solution. They found that the resonance wavelength increases as the NH_3 concentration increases and the sensor with ITO layer of 60 nm gave the best response.

Fiber sensors based on evanescent wave absorption were proposed using bent optical fiber [52] and removed-clad fiber [53]. In [52], NH_3 sensors were proposed using bent optical fibers with PANI and Fe (III) porphyrin-doped PMMA have detection limit in the range of ppm. The cladding of silica MMF was removed using chemical etching and replaced with thin PANI layer (less than 1 μm) in [53] for NH_3 sensing. It was observed that the absorbance spectra increase over certain wavelength (between 500 to 800 nm) as the sensors were exposed to NH_3. However, there is no detailed explanation on the synthesis methods and the type of PANI used in this work. Even though this work is quite dated (2003), but it gave a useful indication that thin PANI layer is a good candidate as an absorbance-based NH_3 sensor.

5. Optical fiber modification, nanomaterials deposition and characterizations

In this section, the development and characterizations of the modified SMF sensing platforms including etched, tapered and etched-tapered platforms will be elaborated. The etching process based on the use of chemical to remove some of the cladding layer. These platforms were characterized in term of output optical power. In the second section, PANI nanostructured thin film preparation and deposition onto the SMF transducing platforms will be highlighted. Finally, the PANI nanofibers fabrication and deposition onto the SMF transducing platforms will be explained. Several micro-characterizations of the fabricated nanostructured thin films were carried out to investigate sensing layer morphology and thickness of the nanostructured thin film. These parameters affect gas sensing performance which will be discussed extensively here.

5.1 Modified SMF fabrication and characterizations

Three types of modified SMF platforms including etched, tapered and etched-tapered platforms were developed and investigated towards ammonia at room temperature. A standard SMF-28 single mode silica fiber (Lucent All-Wave Fiber, 9/125 μm core/cladding diameter ratio) is modified as the optical transducing platforms for NH_3 sensing applications. Each transducing platform was made of 1 m length of SMF. The Tafzel ® polymer jacket enfolding the SMF was removed mechanically over 8 cm by a fiber a stripper. The fabrication processes for the three types of modified SMF platform will be elaborated in the following subsections [2, 11].

5.1.1 Etched SMF sensor

A 48% hydrofluoric acid (HF) (Sigma Aldrich) was used as an agent for chemically etching the SMF. For better holding of the optical fiber platform, both ends of the fiber are fixed using a metal racks so the fiber dangle into the vessel containing the acid used for the etching as shown in **Figure 3**. This is fixed inside a fume hood to prevent direct exposure to HF vapor and creation of aerosols. The etching process was started by filling 100 μl of HF acid in a container using a Pasteur pipette. To fabricate the etched optical fiber transducer, the stripped fiber is fixed as depicted in the figure to control its emersion in HF acid. The fiber is etched in two stages process to control its modification. The first stage includes immersing the stripped area in 48% HF acid at a specific time to produce different etched diameters. After that, the fiber was taken out and cleaned with deionized water for 30 minutes to remove the HF acid residual. The etched fiber is left to dry at room temperature. In the second stage, the fiber was immersed in HF with less concentrations of 12% to reduce the etching rate and hence, more control on the etching dimensions. A variety of parameters affect the etching rate of the cladding such as acid concentration, humidity and temperature. The humidity and temperature are fixed at 67% and 25 C, respectively. As the second stage of etching is finished, the fiber is immersed in deionized water for 30 minutes to remove the remaining of the HF acid as well as to prevent further etching due to residual HF [2, 11].

5.1.2 Tapered SMF sensor

Tapering the SMF is a critical part in the work. Vytran Glass Processing System workstation (GPX-3000, USA) was used to taper transducing platforms. This

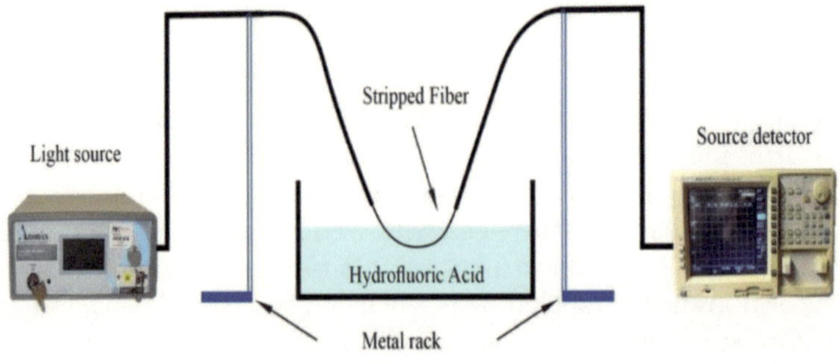

Figure 3.
Setup for etching optical fiber using HF acid [2, 11].

system implies a heating element which is a filament of graphite and two movable fiber holders blocks as a part of tapering control process to generate the recorded profile parameters. The tapered SMF was fabricated using heat-pulling technique which suggests that the fiber diameter changes smoothly and is adiabatically slow as a function of fiber length. Before tapering, the plastic jacket and polymer coating of the fiber are removed approximately 8 cm length. The machine was controlled using a proprietary software on a computer, where the desired tapered fiber profile (waist diameter, waist length, down taper length, and up taper length) can be specified. The waist length, up and down transitions setting for tapering SMF platforms are 10, 2 and 2 mm, respectively. The dimensions of the modified SMF transducing platforms are verified using the CCD camera of Vytran workstation [2, 11].

5.1.3 Fabrication of etched-tapered SMF transducing platform

Fabrication of the etched-tapered SMF (ETSMF) sensing platform was done by combining the two processes; etching and tapering. Firstly, the SMF was etched using HF acid in the same manners described in SubSection 5.1.1 Afterward, the etched SMF was tapered using Vytran workstation according to the proposed settings used in Section 5.1.2 The ETSMF platform is shown in **Figure 4**. Tapering the etched SMF is critical and more challenging than tapering the standard fiber. The reduction in the fiber diameter and the weaken fiber structure due to the etching process increases the difficulty of fragility the modified fiber. Thus, customized holder that fixed the etched fiber during the tapering process was used. The diameters of the optical fibers used in the research fabricated by both etching and tapering were confirmed using the Vytran Glass Processing System workstation GPX- 3000 [11].

In order to verify the compatibility of modified SMF platforms for gas sensing, many experiments were carried out using these sensors coated with PANI nanostructures thin films as sensing layer for NH_3. **Table 1** summarizes the design parameters for the fabricated sensors used in this PhD project. Sensors S1-S4 are ETSMF sensors while Sensors S5 and S6 are the tapered only and etched only sensors. Referring to **Table 1**, it is observed that the core to cladding ratio is found to be varied according to the different modification techniques. It is noted from **Table 1** that in the etched only sensing platform (S6), the original core diameter is unchanged at 9 μm [2, 11]. During the etching process, the cladding layer was dissolved to produce 6 μm thickness diameter. The core to cladding ratio is 0.6. On the other hand, the core diameter is modified as well as the cladding diameter in the case of tapered only sensing platform (S2). The core diameter after tapering became 1.08 μm to give core-cladding ratio as the lowest (0.07) for the tapered only platform that is the lowest relative to other modified fibers [11].

The ETSMF sensing platforms (S1-S4) possess unique core-cladding ratio as compared to the etched only and tapered only platforms. During the etching process, some of the cladding was removed and hence the core becomes more

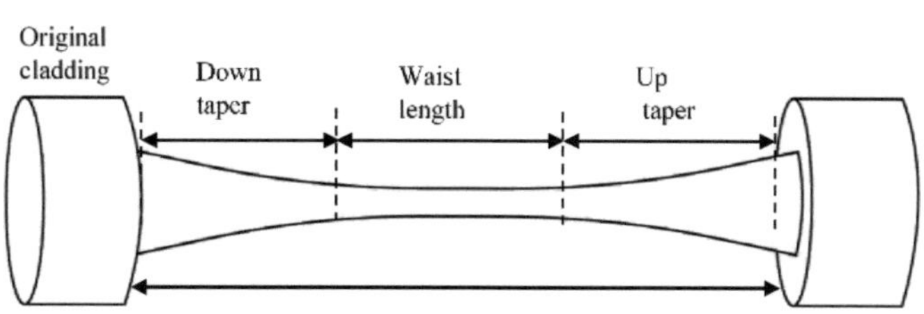

Figure 4.
Etched-tapered sensing platform schematic diagram [2, 11].

Label	Etched fiber diameter (μm)	Tapered fiber diameter (μm)	Fiber core diameter after tapering (μm)	Core to cladding ratio
S1	30	15	4.5	0.3
S2	40	15	2.95	0.2
S3	50	15	2.44	0.16
S4	60	15	2.01	0.15
S5	No	15	1.08	0.07
S6	15	No	9	0.6

Table 1.
Design parameters of the modified SMF sensors [2, 11].

sensitive to the surroundings and the ratio of core-cladding is modified. Afterward, the core diameter is reduced when tapered to the specified settings. The tapering process does not change the core-cladding ratio of the sensing platforms which is equal to the previous etched fiber. For example, S1 sensing platform with 30 μm etched diameter has a core diameter of 9 μm. By tapering process, the cladding and core diameters are changed so that the core diameter is about 4.5 μm core diameter. Consequently, combination of etching and tapering processes yields to a reduction in cladding layer thickness surrounding the core allowing the latter to be more sensitive to the variations in environmental parameters. It is found that the core to cladding ratio for sensor S1 is 0.3 which the highest ratio for the ETSMF platforms [2, 11]. The core cladding ratio for the ETSMF sensing platforms S2-S4 are found to be 0.2 (S2), 0.16 (S3) and 0.15 (S4) as listed in **Table 1**. Stronger response is expected from the ETSMF sensing platforms as a consequence of both surface area and evanescent field enhancement via combination of etching-tapering processes. These three different groups of optical fiber platform were characterized in terms of their optical transmission in terms of the output optical power.

5.2 Characterization of modified SMF transduction platforms

The surface investigation of the unprocessed and modified single mode optical fibers is shown in **Figure 5** based on scanning electron microscopy (SEM - Hitachi SU1510, Japan). The unprocessed single mode optical fiber with a diameter of 125 μm is shown in **Figure 5(a)**. **Figure 5b** depicts the etched SMF using hydrofluoric acid with total fiber diameter of 9.7 μm. **Figure 5(c)** and **(d)** shows the etched-tapered with a diameter of 77.7 μm after etching and waist diameter of 15 μm after tapering. The tapered part of the etched-tapered fiber is shown in **Figure 5(c)**. The fiber exhibited a uniform transition with surface roughness. The downward transition of the etched-tapered SMF with 15 μm waist diameter is shown in **Figure 5(d)**. it can be noted that the unprocessed SMF has a smooth surface. On the other hand, the etched fiber exhibited a rough surface due to its processing with hydrofluoric acid. The surface roughness of the modified SMF is superior to enhance the surface area of sensing layer deposited onto it which allows stronger interaction between sensing layer molecules and the gas molecules [2, 11].

5.3 Synthesis, deposition and characterization of polyaniline (PANI) nanofiber thin film sensing layer

PANI nanofiber was deposited on modified SMF transducing platforms as sensing layers towards NH_3. In the next subsections, the synthesis and

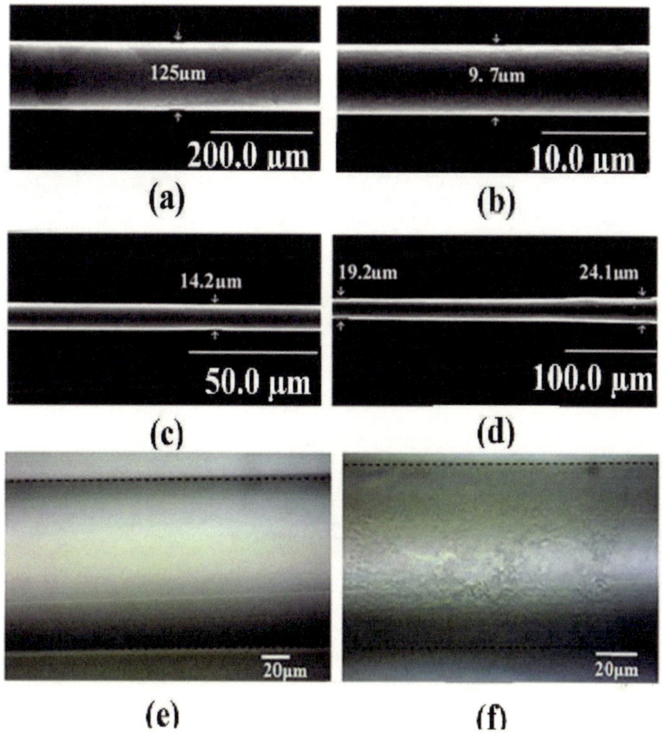

Figure 5.
SEM images of modified SMF platforms, (a) original SMF, (b) the etched SMF, (c) the etched-tapered SMF and (d) the down ward transition for etched-tapered SMF in (c) and microscopic images of modified SMF platforms (e) original SMF and (f) a 77.7 µm etched SMF [2, 11].

micro-characterization of these nanostructured thin films will be presented. The modified SMF transducers were coated with PANI nanofiber thin films as a sensing layer towards NH_3. PANI solution was prepared by dissolving and dispersing a 15 mg PANI powder with 15 mg camphor sulfonic acid (Sigma Aldrich) in 8 ml of chloroform ($CHCl_3$). This resulted in CSA-doped PANI nanofiber of green color solution with concentrations of 3.75 mg/ml [11]. The camphor sulfonic acid was implied to boost the ability of solving PANI in chloroform. The resulted solution was stirred for 1 hour using magnetic bar.

To generate a homogeneous solution to be used to deposited on the SMF transducers and glass substrates, the solvent was sonicated for 1 hour at room temperature using Hielcher, Ultrasound Technique, UPS2005 ultra sound processor. The fabricated PANI-CSA solution had dark green color. This indicates the doping with CSA was carried out successfully. Glass is chosen as one of the substrates for PANI micro-characterization. The optical fibers and glass substrates were heated up to 50°C for 30 minutes prior to deposition of PANI using hotplate. The heating is important to increase the binding of the nanomaterial and generating uniform films. The prepared samples were left to dry for 1 hour at room temperature. The coating process was done using a fume hood [2, 11].

The polyaniline coated on the glass substrate was investigated using scanning electron microscopy as described in **Figure 6(a)** to prove its morphology. As can be noted from the figure, PANI deposited on the glass exhibits a random distribution over the substrate surface in cluster forms with different sizes. As can be noted from the **Figure 6(b)**, the non-uniform nanofibers agglomerated to produce the cluster morphology. These PANI nanofibers exhibits a typical length of 2.5–3.5 µm with diameter in the range of 180–200 nm. **Figure 6(c)** presents the image of the PANI

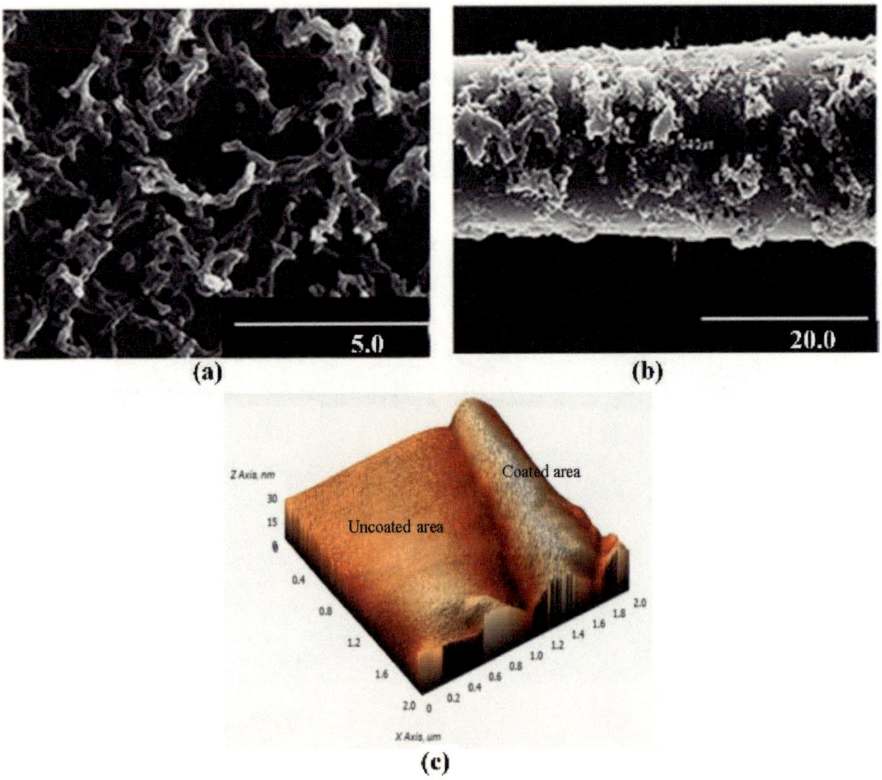

Figure 6.
SEM images of (a) PANI nanofibers on glass, (b) etched-tapered SMF transducer coated with PANI nanofibers and (c) 3D AFM image of uncoated and PANI coated areas on a glass substrate [2, 11].

thin film taken with the aid of an atomic force microscope (NT-MDT Solver NEXT AFM). The average thickness of PANI thin film was found to be 400 nm while the its surface roughness was approximately 228.2 nm [11]. Surface roughness is significant in the applications of gas sensing as it enhances the surface area which rises the active interaction sites between the gas molecules and the sensing layer. Consequently, increases the sensor sensitivity [11].

5.4 Ammonia sensing performance of modified SMF coated with PANI nanofiber

The setup used to investigate the optical response of the modified SMF sensors towards NH_3 is outlined in **Figure 7**. This setup is used to prove the behavior of the sensors in the visible wavelength range (600–750 nm) and C-band wavelength range (1535–1565 nm). Based on the setup, the modified SMF sensor is placed inside a gas chamber which contains a gas inlet and outlet as well as FC/PC connecting adapters to fit the sensors. The sensor is connected to a light source (tungsten-halogen lamp (Ocean Optics HL2000) for visible and C-band (Ammonics) for C-band) and the other end is connected to a detection system spectrophotometer (Ocean Optics USB4000) for visible and OSA for C-band). A proprietary software is used to record and measure the responses from the detection system. A gas calibration system (AALBORG) is deployed to vary the gas concentrations and purging time, automatically. NH_3 of 1% concentrations in 99% synthetic air is purged into the chamber via the MFCs. Another pure synthetic air is used as the reference gas. The gas flow is fixed at a rate of 200 sccm. This is completely

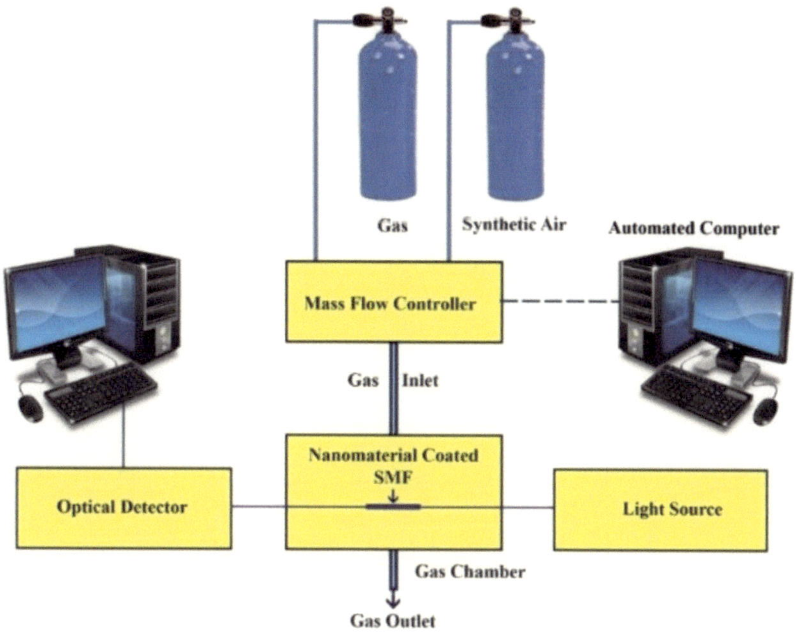

Figure 7.
Experimental setup for modified SMF ammonia sensing coated with PANI thin film [2, 11].

computerized using Labview control program. Certified NH_3 and synthetic air gas cylinders (Linde, Malaysia-Singapore Sdn Bhd) were used in the mixing and purging of the gases into the chamber. The mixing was done for the purpose of changing the concentration of ammonia in the range of 0.125–1%. The dynamic response of the SMF sensors was investigated upon exposure to NH_3 with different concentrations at room temperature. This was carried out through measuring cumulative absorbance of the sensor as it exposed to a NH_3 at abovementioned concentrations. Each gas concentration cycle was persisted for 8 minutes while the sensor air regeneration lasted for 15 minutes [2, 11].

In order to verify the compatibility of modified SMF sensors for gas sensing, many experiments were carried out using these sensors towards NH_3. The sensors were coated with sprayed PANI nanostructure thin films as a sensing layer. The sensing performances were investigated and analyzed in both visible and C-band wavelength ranges. The details of morphology and thickness the of the PANI sensing layer was introduced in Section 5.3. Different modified SMF, namely etched, tapered and ETSMFs were investigated towards NH_3 gas with different concentrations. The design parameters for the fabricated sensors used are summarized in **Table 1**. These platforms' dynamic response was investigated towards NH_3 in the C band wavelength range [11].

Figure 8 demonstrates the dynamic responses of the etched-tapered SMF (S1-S4) while **Figure 9** represents that of tapered and etched SMF (S5 and S6) sensors against different concentrations of ammonia, respectively. SMF sensors with different modifications proved proportional increase in the output optical power against NH_3 concentrations. The etched-tapered sensors exhibited superior response magnitude over that of the sensors with other modifications. The etched-tapered sensors (S1-S4) shows response of 1.6, 1.5, 1.39, and 1.29 dBm, respectively, when exposed to 1% of ammonia while the tapered and etched sensors response is 0.84 and 0.68 dBm, respectively. Lower increases are noted at lower NH_3 concentrations. As exposed to 0.125% NH_3 concentration, the modified sensors (S1-S6) exhibited a change

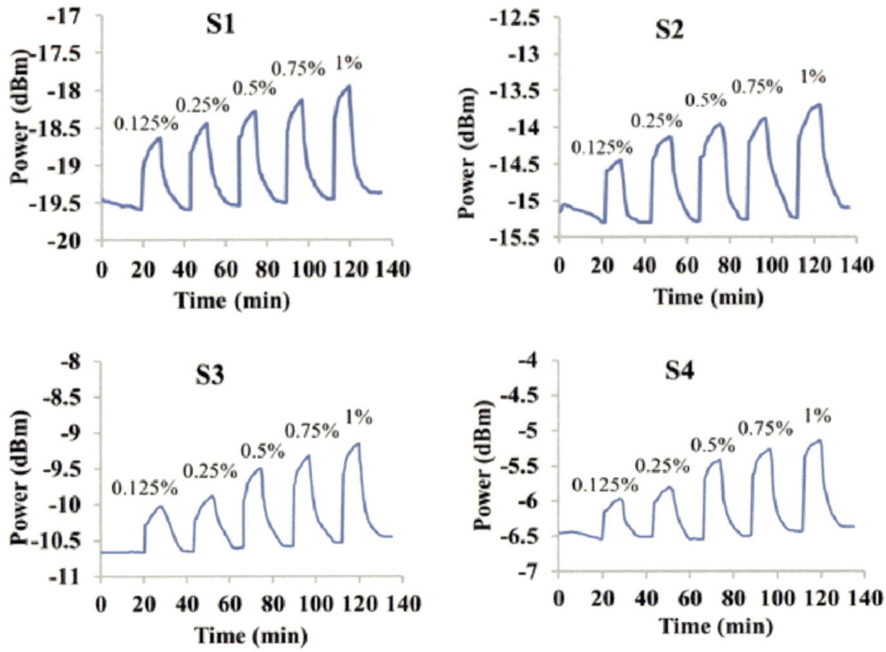

Figure 8.
Dynamic response of the etched-tapered SMF sensors (S1-S4) exposed to different NH_3 concentrations in the C-band wavelength range [2].

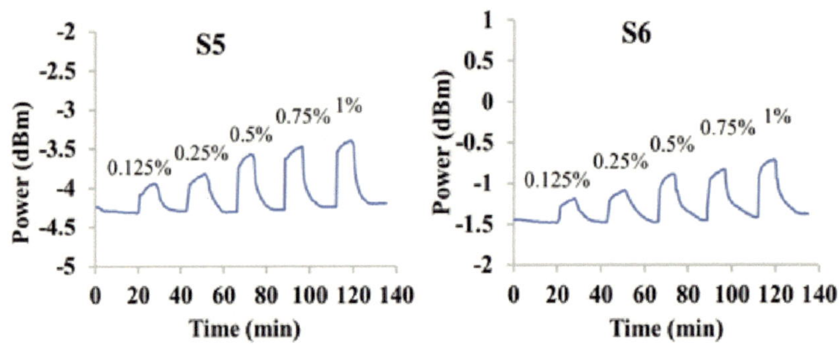

Figure 9.
Dynamic responses for tapered and etched SMF sensors (S5 and S6) when exposed to NH_3 with different concentrations [2].

of 0.96, 0.86, 0.68, 0.55, 0.36, and 0.29 dBm, respectively [2, 11]. Generally, SMF sensors with different modifications proved different responses and recovery times are. Moreover, the response time is inversely proportional to the ammonia concentrations while the recovery time is linearly proportional to it for all SMF sensors with different modifications investigated in the C-band range. The core-to-cladding ratio due to etching and tapering processes of the modified SMF sensors affects the response and recovery times as shown in **Figures 8** and **9** [2, 11].

The average response time for the modified sensors (S1-S6) against ammonia are 58–71 s, 79 s, and 92 s, respectively. The average recovery time designates the opposite performance for the sensors with different modification techniques. The values for modified sensors (S1-S6) against ammonia are 466–453 s, 380 s, and 360 s, respectively. Different sensing performances for different types of modified SMF sensors is attributed to different rates of ammonia molecules adsorption/

desorption on each kind of the sensors surfaces. For example, fastest response but slowest recovery is observed for the etched-tapered sensors. Contrary, the etched-only sensor shows the reverse time responses. The response of the modified SMF sensors presented here and investigated in the C-band wavelength range were significantly improved compared with the tapered MMF sensors described in a previous study investigated in the visible wavelength range [54]. The response time of etched-tapered SMF sensors (62 s) was more than twice that of the tapered MMF sensors (2.27 minutes or 136.2 s) [11, 54]. The recovery time of the SMF and the MMF sensors were 453 and 583.8 s, respectively [54]. The sensor introduced by Airoudj et al. [55] investigated in the visible to near-infrared wavelengths (632.8–980 nm) based on the single mode planar polymer waveguide coated with PANI exhibited response and recovery times of 180 s and 480 s, respectively [2, 11].

The normalized cumulative Δ optical power of the modified SMF sensors coated with PANI thin film against ammonia is depicted in **Figure 10**. The etched-tapered SMF sensors (S1–S4) proved significant enhancement in response compared with other modifications. On the contrary, the etched-only sensor exhibited the lowest response. The etched-tapered SMF response as exposed to ammonia is improved when the SMF cladding thickness is reduced due to etching process. For example, sensor (S1) with smallest etching diameter shows the strongest response among the etched-tapered sensors. This result principally is a consequence of the strong evanescent field energy propagating out of the core physical dimensions for the modified optical fiber sensors into the sensing layer of PANI. Additionally, the modified SMF sensors integrated with PANI show a variation in response that is proportional to the ammonia concentrations. The normalized cumulative optical powers for the sensors (S1–S6) are 17.7%, 14.6%, 10.8%, 9.5%, 6.3%, and 1.9% at the limit of detection of the sensors [2, 11]. The practical of the gas system used in the work, the limit of detection for the SMF sensors is found to be 0.04% or 400 ppm. Based on established technique introduced by Mola et al. [11, 56], the limit of detection for the etched-tapered SMF Sensor S1 is 0.0025%, which is equal to 25 ppm. Accordingly, the established PANI coated SMF sensor can detect ammonia gas concentration below the ammonia lowest tolerable exposure limit reported by OSHA [11].

The sensitivities for the SMF sensors (S1-S6) are found to be 231.5 (S1), 209.7 (S2), 172.1 (S3), 146.6 (S4), 100.4 (S5) and 81.2 (S6). The ETSMF sensors (S1-S4) proved considerably higher sensitivity towards NH_3 compared to the tapered only (S5) and etched only sensors (S6). The sensitivity of the ETSMF sensor (S1-S4) are 2.8, 2.3,1.7 and 1.5 times the sensitivity of the tapered only SMF sensor (S5)

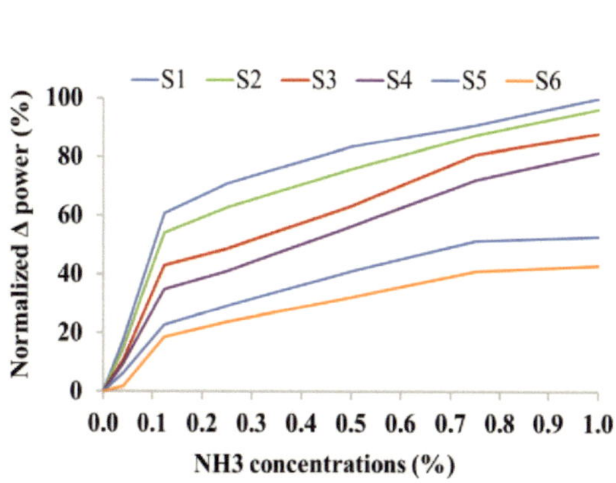

Figure 10.
Normalized Δ power versus NH_3 concentrations for the modified SMF sensors (S1-S6) in the C-band wavelengths range [2, 11].

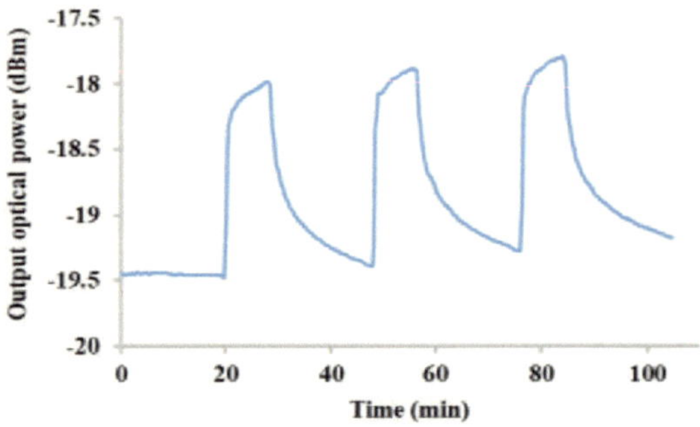

Figure 11.
Repeatability for the ETSMF sensor S1 coated with PANI nanostructured thin film as exposed to 1% NH_3 for three cycles [2, 11].

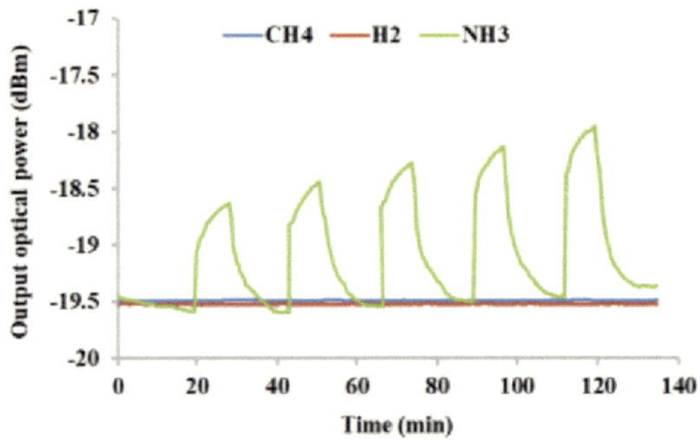

Figure 12.
Optical response for ETSMF sensor S1 coated with PANI nanostructured thin film towards CH_4, H_2 and NH_3 [2].

and 9.3, 7.7, 5.7 and 5 times that of the etched only SMF sensor (S6) [2, 11]. Based on **Figure 10**, the sensitivities for the SMF sensors (S1–S6) are found to be 231.5, 209.7, 172.1, 146.6, 100.4, and 81.2, respectively [2, 11]. As compared with the tapered and etched sensors, the etched-tapered SMF sensors have higher sensitivity towards NH_3 [2].

Figure 11 shows The repeatability and reversibility of etched-tapered Sensor S1 against 1% ammonia concentration for three cycles each lasts for 8 minutes of ammonia, followed by purging purified air for 15 minutes. These three cycles in the figure exhibits slight difference when exposed towards 1% ammonia. Furthermore, the base time experienced slight shift as a result of incomplete elimination of ammonia from the PANI sensing layer when the air was purged for 15 min. The etched-tapered SMF sensor shows good repeatability and reversibility as demonstrated in the figure.

The etched-tapered Sensor S1 was examined towards methane (CH_4), hydrogen (H_2), and ammonia (NH_3) to prove its selectivity towards ammonia. The range of concentrations of these gases utilized in this test was from 0.125–1% at room temperature. The sensor exhibits superior response towards ammonia compared to that due to methane and hydrogen as depicted in **Figure 12**. Accordingly, the sensor can be described to be highly selective towards NH_3.

6. Conclusion

Simple and low-cost modified SMF platforms were successfully designed, developed and investigated for optical sensing towards NH_3 with low concentrations.

The modification processes performed on the SMFs were etching, tapering and combination of etching-tapering. The sensing performance of the modified SMF sensors coated with PANI nanofibers thin films was investigated towards NH_3 at room temperature for C-band wavelengths range (1535–1565 nm). The investigation showed that the principle of the gas sensors used is the change in the light characteristics because of the interaction between the NH_3 molecules and PANI sensing layer. The interaction between the NH_3 molecules and the PANI sensing layer reduces the absorbance in the C-band range and inversely proportional to the NH_3 concentrations. Consequently, the transmitted optical power increased. These PANI nanofiber thin films show high sensitivity towards NH_3 with low concentration as well as good repeatability indication. The performance of the modified SMF sensors was found to be dependent on the modification technique used in the fabrication of the SMF platform as well as the thickness of the cladding layer after modification (core/cladding ratio). The investigation on the NH_3 optical sensing performance using absorbance measurement proved that the ETSMF showed superior response than the other modified fibers and thus, highly potential for novel optical transducer. The ability of the ETSMF coated with PANI thin films operates at room temperature indicates its promising candidate for NH_3 sensing applications such as chemical plant leakage remote sensor. Particularly, its response in the C-band wavelength range allows easy integration with the existing all optical fiber communication networks infrastructures. The modification technique used in the fabrication of the SMF platform namely etching, tapering or combination of both strongly affects the performance of the modified SMF sensors. Furthermore, the performance of the modified SMF sensors is also dependent on the thickness of the cladding layer after modification and core/cladding ratio. The investigation on the ammonia optical sensing performance demonstrated that the response of etched-tapered SMF is the stronger among that of other modified sensors. Thus, the developed etched-tapered SMF sensors show high potential to be a novel optical transducer. The ability of the etched-tapered SMF coated with PANI to perform at room temperature makes it a good candidate for ammonia sensing and remote monitoring applications. Particularly, its strong response in the C-band wavelength range makes it is easy to be integrated with all well-established optical fiber communication network infrastructures such as fiber to the home. The developed sensor exhibits good repeatability and reversibility. The limit of detection of the modified SMF sensor was 0.0025% (25 ppm).

Author details

Husam Abduldaem Mohammed[1]* and Mohd Hanif Yaacob[2]

1 Electronic and Communication Engineering Department, College of Engineering, University of Baghdad, Baghdad, Iraq

2 Wireless and Photonics Network Research Centre, Faculty of Engineering, Universiti Putra Malaysia, Serdang, Selangor, Malaysia

*Address all correspondence to: husam.a@coeng.uobaghdad.edu.iq

IntechOpen

© 2020 The Author(s). Licensee IntechOpen. This chapter is distributed under the terms of the Creative Commons Attribution License (http://creativecommons.org/licenses/by/3.0), which permits unrestricted use, distribution, and reproduction in any medium, provided the original work is properly cited.

References

[1] H. A. Mohammed, S. A. Rashid, M. H. Abu Bakar, S. B. Ahmad Anas, M. A. Mahdi and M. H. Yaacob, "Fabrication and Characterizations of a Novel Etched-tapered Single Mode Optical Fiber Ammonia Sensors Integrating PANI/GNF Nanocomposite," *Sensors and Actuators B: Chemical,* vol. 287, pp. 71-77, 2019/05/15/ 2019.

[2] H. A. Mohammed, N. A. Rahman, M. Z. Ahmad, M. H. A. Bakar, S. B. A. Anas, M. A. Mahdi and M. H. Yaacob, "Sensing performance of Modified Single Mode Optical Fiber Coated with Nanomaterials Based Ammonia Sensors Operated in the C-Band," *IEEE Access,* vol. 7, pp. 5467-5476, 2019.

[3] S. G. a. S. Albin, "Transmission Property and evanescent wave absorption of cladded multimode fiber tapers," *Opt. Express,* vol. 11, pp. 215-223, 2003.

[4] M. Y. J. Dai, Y. Chen, K. Cao, H. Liao, and P. Zhang, "Side-polished fiber Bragg grating hydrogen sensor with WO3-Pd composite film as sensing materials," *Opt. Express,* vol. 19, no. 7, pp. 6141-8, Mar. 2011.

[5] S. W. J. J. C. Hsu, and Y. S. Sun, "Simulation and experiments for optimizing the sensitivity of curved D-type optical fiber sensor with a wide dynamic range," *Opt. Commun.,* vol. 341, pp. 210-217, 2015.

[6] D. Engles, S. Prashar, A. Singh and M. T. Student, "Etched FBG as chemical sensor for fuel adultration," *Int. J. Eng. Res. Technol.,* vol. 1, no. 4, pp. 1-5, 2012.

[7] J. Dai, M. Yang, X. Yu, K. Cao and J. Liao, "Greatly etched fiber Bragg grating hydrogen sensor with Pd/Ni composite film as sensing material," *Sensors and Actuators B: Chemical,* vol. 174, pp. 253-257, 2012.

[8] Y. Tian, W. Wang, N. Wu, X. Zou and X. Wang, "Tapered optical fiber sensor for label-free detection of biomolecules," *Sensors (Basel).* vol. 11, no. 4, pp. 3780-90, Jan. 2011.

[9] H. L. M. I. Zibaii, Z. Saeedian, and Z. Chenari, "Nonadiabatic tapered optical fiber sensor for measurement of antimicrobial activity of silver nanoparticles against *Escherichia coli*," *J. Photochem. Photobiol. B Biol.,* vol. 135, pp. 55-64, 2014.

[10] D. Liu, Q. Sun, P. Lu, L. Xia and *C. sima*, "Research Progress in the Key Device and Technology for Fiber Optic Sensor Network," *Photonic Sensors,* vol. 6, pp. 1-25, 2016.

[11] H. A. Mohammed, "OPTICAL CODE DIVISION MULTIPLE ACCESS BASED GAS SENSOR NETWORK USING MODIFIED SINGLE MODE FIBER COATED WITH POLYANILINE/GRAPHITE NANOFIBER," Doctor of Philosophy, Department of Computer and Communications Engineering, Universiti Putra Malaysia (UPM) Malaysia, 2018.

[12] J. M. Tam, S. Szunerits and D. R. Walt, "Optical Fibers for Nanodevices," *Encyclopedia of Nanoscience and Nanotechnology Edited by H. S. Nalwa,* vol. 8, pp. 167-177, 2004.

[13] W. Jin, H. L. Ho, Y. C. Cao, a. J. Ju and L. F. Qi, "Gas detection with micro- and nano-engineered optical fibers," *Opt. Fiber Technol.,* vol. 19, no. , no. 6 PART B., p. 19, 2013.

[14] S. Xue, M. A. van Eijkelenborg, G. W. Barton and P. Hambley, "Theoretical, numerical, and experimental analysis of optical fiber tapering," *Journal of Lightwave Technology,* vol. 25, no. 5, pp. 1169-1176, 2007/05/01 2007.

[15] R. Sivacoumar, M. Vinoth and Z. C. Alex, "Tapered optical fiber

bioSensor for testosterone detection," *Tagungsband,* pp. 821-825, 2012.

[16] W. B. Ji, H. H. Liu, S. C. Tjin, K. K. Chow and A. Lim, "Ultrahigh sensitivity refractive index sensor based on optical microfiber," *IEEE Photonics Technology Letters,* vol. 24, no. 20, pp. 1872-1874, 2012.

[17] S. A. B. Ibrahim, "Tapered optical fiber coated With polyaniline nanostructures for ammonia sensing " Doctor of Philosophy, Universiti Putra Malaysia, 2016.

[18] T. Chen, H. Chen, C. Hsu, C. Huang, C. Chang, P. Chou and W. Liu, "On an ammonia gas sensor based on a Pt/AlGaN heterostructure field-effect transistor," *IEEE Electron Device Lett,* vol. 33, no. 4, April 2012 2012.

[19] P.-C. Chou, H.-I. Chen, I.-P. Liu, C.-C. Chen, J.-K. Liou, K.-S. Hsu and W.-C. Liu, " On the ammonia gas sensing performance of a RF sputtered NiO thin-film sensor," *IEEE Sensors Journal,* vol. 15, no. 7, p. 5, July 2015 2015.

[20] S. G. Pawar, M. A. Chougule, S. L. Patil, B. T. Raut, P. R. Godse, S. Sen and V. B. Patil, "Room temperature ammonia gas sensor based on polyaniline-TiO2 nanocomposite," *IEEE Sensors J.,* vol. 11, no. 12, p. 7, Dec. 2011 2011.

[21] S. A. Ibrahim, N. A. Rahman, M. H. Abu Bakar, S. H. Girei, M. H. Yaacob, H. Ahmad and M. A. Mahdi, "Room temperature ammonia sensing using tapered multimode fiber coated with polyaniline nanofibers," *Opt Express,* vol. 23, no. 3, pp. 2837-45, Feb 9 2015.

[22] G. K. Mani and J. B. B. Rayappan, "A highly selective room temperature ammonia sensor using spraydeposited zinc oxide thin film," *Sensors and Actuators B,* vol. 183, p. 8, 2013.

[23] Praxair, "Ammonia , anhydrous ammonia , anhydrous safety data sheet P-4562, pp. 1-9,," ed, 2015.

[24] B. Timmer, W. Olthuis and v. d. Berg, "Ammonia Sensors and Their Applications-a Review," *Sens. Actuators B, Chem,* vol. 107, p. 12, 2005.

[25] K. Inus, "2 die in ammonia leak at plant," in *New Straits Times*, ed. Malaysia, 2016, p. 25.

[26] B. Timmer, W. Olthuis and A. v. d. Berg, "Ammonia sensors and their applications—a review," *Sensors and Actuators B: Chemical,* vol. 107, no. 2, pp. 666-677, 2005.

[27] A. Sutti, C. Baratto, G. Calestani, C. Dionigi, M. Ferroni, G. Faglia and G. Sberveglieri, "Inverse opal gas sensors: Zn(II)-doped Tin dioxide systems for low temperature detection of pollutant gases," *Sens. Actuators B,* vol. 130, p. 7, 2008.

[28] X. Xu, X. Fang, H. Zeng, T. Zhai, B. Y. and D. Golberg, "One-dimensional nanostructures in porous anodic alumina membranes," *Sci. Adv. Mater,* vol. 2, p. 22, 2010.

[29] J. Wang, P. Yang and X. Wei, "High-performance, room-temperature, and no-humidity-impact ammonia sensor based on heterogeneous nickel oxide and zinc oxide nanocrystals," *ACS Appl Mater Interfaces,* vol. 7, no. 6, pp. 3816-24, Feb 18 2015.

[30] J. H. W. Gopel and J. N. Zemel, *Sensors: A comprehensive survey*. New York: VCH, 1991.

[31] N. Yamazoe, "Toward innovations of gas sensor technology," *Sensors and Actuators B: Chemical,* vol. 108, no. 1, pp. 2-14, 2005.

[32] E. Della Gaspera, et al., "Comparison study of conductometric,

optical and SAW gas sensors based on porous sol–gel silica films doped with NiO and Au nanocrystals," *Sensors and Actuators B: Chemical,* vol. 143, no. 2, pp. 567-573, 2010.

[33] J. L. Bredas and G. B. Street, "Polarons, bipolarons, and solitons in conducting polymers," *Acc. Chem. Res,* vol. 18, no. 10, pp. 309-315, 1985.

[34] Chuanjun Liu and K. Hayashi, "A Gold Nanoparticle/Polyaniline Nanofiber Sensor for Detecting H2S Impurity in Hydrogen Fuel," *Extended Abstracts of the 2013 International Conference on Solid State Devices and Materials, Fukuoka, 2013,,* pp. 412-413, 2013.

[35] Z. Jin, Y. Su and Y. Duan, "Development of a polyaniline-based optical ammonia sensor," *Sensors and Actuators B: Chemical,* vol. 72, no. 1, pp. 75-79, 1/5/ 2001.

[36] D. Nicolas-Debarnot and F. Poncin-Epaillard, "Polyaniline as a new sensitive layer for gas sensors," *Anal. Chim. Acta* vol. 475, no. 1-2, p. 16, 2003 2003.

[37] G. G. Wallace, P. R. Teasdale, G. M. Spinks and L. A. Kane-Maguire, *Conductive electroactive polymers: intelligent polymer systems*. CRC press, 2009.

[38] A. L. Sharma, K. Kumar and A. Deep, "Nanostructured polyaniline films on silicon for sensitive sensing of ammonia," *Sensors and Actuators A: Physical,* vol. 198, pp. 107-112, 2013.

[39] A. D. Aguilar, E. S. Forzani, L. A. Nagahara, I. Amlani, R. Tsui and N. J. Tao, "A breath ammonia sensor based on conducting polymer nanojunctions," *IEEE Sens. J.,* vol. 8, no. 3, p. 5, 2008.

[40] H.-W. Zan, W.-W. Tsai, Y. Y.-R. Lo, M. Wu and Y.-S. Yang, "Pentacene-based organic thin film transistors for ammonia sensing," *IEEE Sensors Journal,* vol. 12, no. 3, p. 8, Mar. 2012 2012

[41] T.-Y. Chen, H.-I. Chen, Y.-J. Liu, C.-C. Huang, C.-S. Hsu, C.-F. Chang and W.-C. Liu, "Ammonia sensing characteristics of a Pt/AlGaN/GaN Schottky diode," *Sensors and Actuators B: Chemical,* vol. 155, no. 1, pp. 347-350, 7/5/ 2011.

[42] Y.-S. Lee, "Visible optical sensing of ammonia based on polyaniline film," *Sensors and Actuators B: Chemical,* vol. 93, no. 1, p. 5, 2003.

[43] Z. Jin, Y. Su and Y. Duan, "Development of a polyaniline-based optical ammonia sensor " *Sensors and Actuators B: Chemical,* vol. 72, no. 1, p. 5, 2001.

[44] S. Virji, "Polyaniline Nanofiber Gas Sensors Examination of Response Mechanisms," vol. 4, no. 3, p. 6, 2004.

[45] N. A. Rahman, "Electrospun conducting polymer nanofibers for biomedical applications," The University of Auckland, New Zealand, 2012.

[46] H. Bai and G. Shi, "Gas Sensors Based on Conducting Polymers," *Sensors (Basel, Switzerland),* vol. 7, no. 3, pp. 267-307, 03/07 10/30/received 03/02/ accepted 2007.

[47] W. Guiqiang, X. Wei and Z. Shuping, "The production of polyaniline/graphene hybrids for use as a counter electrode in dye-sensitized solar cells," *Electrochimica Acta,* vol. 66, pp. 151-157, 2012.

[48] A. M. Lentz, G. Gheno, T. Maraschin, J. A. Malmonge, N. R. de Souza Basso, N. M. Balzaretti, . . . G. B. Galland, "Nanocomposites of polyethylene/polyaniline/graphite with special morphology," *Polymer Composites,* 2017.

[49] H. Kebiche, D. Debarnot, A. Merzouki, F. Poncin-Epaillard and N. Haddaoui, "Relationship between ammonia sensing properties of polyaniline nanostructures and their deposition and synthesis methods," *Analytica Chimica Acta,* vol. 737, pp. 64-71, 8/6/ 2012.

[50] L. Ai, J. C. Mau, W. F. Liu, T. C. Chen and W. K. Su, "A volatile-solvent gas fiber sensor based on polyaniline film coated on superstructure fiber Bragg gratings," *Measurement Science and Technology,* vol. 19, no. 1, p. 017002, 2008.

[51] S. K. Mishra, D. Kumari and B. D. Gupta, "Surface plasmon resonance based fiber optic ammonia gas sensor using ITO and polyaniline," *Sensors and Actuators B: Chemical,* vol. 171-172, pp. 976-983, 2012.

[52] L. W. Y. Huang, and S. Tao, "Development and evaluation of optical fiber NH3 sensors for application in air quality monitoring," *Atmos. Environ.,* vol. 66, pp. 1-7, 2013.

[53] Y. Jianming and M. A. El-Sherif, "Fiber-optic chemical sensor using polyaniline as modified cladding material," *Sensors Journal, IEEE,* vol. 3, no. 1, pp. 5-12, 2003.

[54] S. A. Ibrahim, N. A. Rahman, M. H. Abu Bakar, S. H. Girei, M. H. Yaacob, H. Ahmad and M. A. Mahdi, "Room temperature ammonia sensing using tapered multimode fiber coated with polyaniline nanofibers," *Optics Express,* vol. 23, no. 3, pp. 2837-2845, 2015/02/09 2015.

[55] A. Airoudj, D. Debarnot, B. Bêche and F. Poncin-Epaillard, "A new evanescent wave ammonia sensor based on polyaniline composite," *Talanta,* vol. 76, no. 2, pp. 314-319, 7/15/ 2008.

[56] H. A. Molla, R. Bhowmick, A. Katarkar, K. Chaudhuri, S. Gangopadhyay and M. Ali, "A novel rhodamine-3,4-dihydro-2H-1,3-benzoxazine conjugate as a highly sensitive and selective chemosensor for Fe3+ ions with cytoplasmic cell imaging possibilities," *Analytical Methods,* vol. 7, no. 12, pp. 5149-5156, 2015.

Chapter 5

Fiber Bragg Grating Sensors Integration in Fiber Optical Systems

Janis Braunfelds, Sandis Spolitis, Jurgis Porins and Vjaceslavs Bobrovs

Abstract

Fiber Bragg grating (FBG) sensors are a progressive passive optical components, and used for temperature, strain, water level, humidity, etc. monitoring. FBG sensors network can be integrated into existing optical fiber network infrastructure and realized structural health monitoring of roads, bridges, buildings, etc. In this chapter, the FBG sensor network integration in a single-channel and multichannel spectrum sliced wavelength division multiplexed passive optical network (SS-WDM-PON) is presented and assessed. The operation of both the sensors and data transmission system, over a shared optical distribution network (ODN), is a challenging task and should be evaluated to provide stable, high-performance mixed systems in the future. Therefore, we have investigated the influence of FBG temperature sensors on 10 Gbit/s non-return-to-zero on–off keying (NRZ-OOK) modulated data channels optical transmission system. Results show that the crosstalk between both systems is negligible. The successful operation of both systems (with BER $< 2 \times 10^{-3}$ for communication system) can be achieved over ODN distances up to 40 km.

Keywords: fiber Bragg grating (FBG) sensors, sensor network, WDM-PON, SS-WDM-PON

1. Introduction

Optical fiber sensors are classified as intensity, phase, polarization, and wavelength modulated sensors based on their operating principles [1]. Fiber Bragg grating (FBG) sensors are wavelength modulated sensors or sensors which detect physical parameter (strain, temperature, and others) based on wavelength changes. Fiber Bragg grating (FBG) technology typically is used in optical filters, dispersion compensation modules, and sensors solutions. FBG sensors are passive optical components with high sensitivity and immune to electromagnetic interference and radio frequency interference, which can be integrated into existing optical fiber network infrastructure for structural health monitoring (SHM) applications [2–4]. FBG sensors can be used for roads, tunnels, bridges, rails, aircrafts SHM, as well as civil engineering, security, oil, and gas solutions monitoring [5]. With FBG sensors, it is possible to monitor various physical parameters such as temperature, strain, vibration, pressure, humidity, etc. [6–8].

It is necessary to analyze sensor influence on deployed and operating fiber optical communications systems data channels before FBG sensors integration in this fiber optical network infrastructure. Optical sensors signal interrogation (OSSI) units maximal monitoring distance between monitoring equipment and FBG sensors can be longer than 40 km.

First, in this paper optical sensor and single–channel 10 Gbit/s transmission system compatibility and co-operation were experimentally evaluated in the fiber-optical transmission system (FOTS) laboratory of Riga Technical University, Communication Technologies Research Center (RTU SSTIC), as described in Section 2 of this article.

Further, we have also demonstrated the collaboration with 32-channel spectrum-sliced wavelength-division-multiplexing passive optical network (SS-WDM PON) data channels and FBG sensor network in the simulation environment, as described in Section 3 of this article. Results showed that the optical transmission system with SS-WDM PON data and FBG sensor channels is an energy and cost-efficient solution, because its transmitter part is realized using a single amplified spontaneous emission (ASE) light source.

2. Evaluation of compatibility and co-operation on fiber-optic FBG sensor and single channel 10 Gbit/s NRZ-OOK transmission system

In this section, FBG temperature sensor integration and co-operation with operating fiber-optical transmission system are experimentally evaluated in the laboratory environment. The transmitter part of the experimental setup (see **Figure 1**) includes a broadband light source - ASE source, which is necessary to provide the operation of the deployed FBG sensor.

The measured output spectrum of the ASE light source is shown in **Figure 2**. The maximal peak power of around −10 dBm is located in wavelength bands of 1532–1534 and 1550–1560 nm. The high output power of the ASE light source and FBG sensor reflectivity is essential when monitoring distance (between the OSSI unit and FBG sensor) is long. Broad spectral band and fixed output power of the light source spectrum are crucial for multiplexing many sensors.

The output of the ASE light source is connected with an optical bandpass filter (OBPF). An OBPF (wavelength range: 1530 to 1610 nm (C&L Band), crosstalk >50 dB, bandwidth: 0.2 to 10.0 nm) is used to filtered spectral band for FBG temperature sensor. The spectral band is calculated based on FBG sensor defined operating

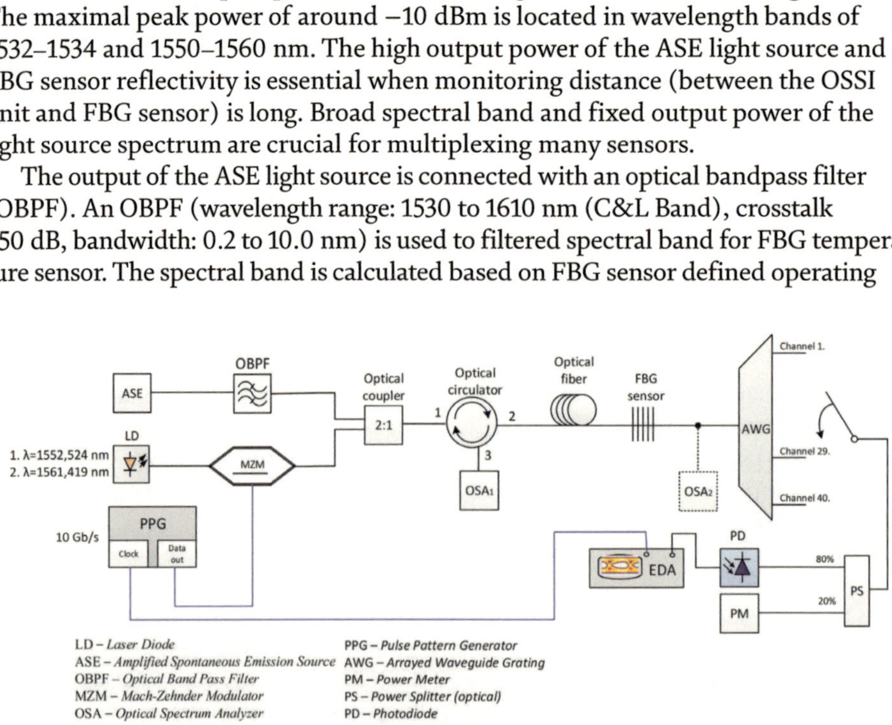

Figure 1.
Experimental setup of single–channel 10 Gbit/s transmission system with integrated optical FBG sensor.

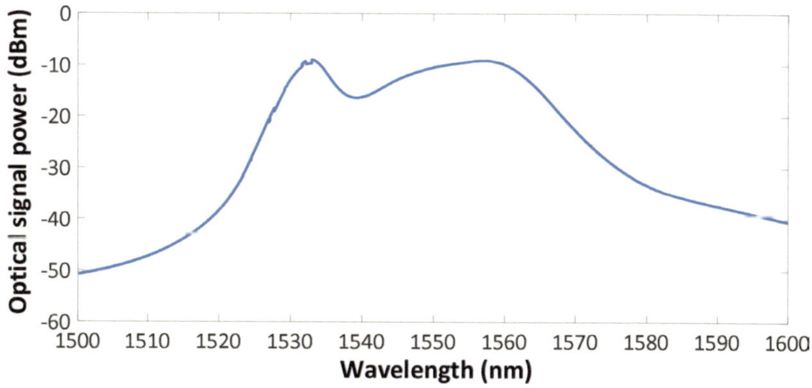

Figure 2.
Measured ASE output spectrum.

temperature band (−20 to +40°C). Temperature change by one degree causes a wavelength shift of 10.174 pm, taking into account the FBG reference temperature of 26°C ($\lambda_{ref} = 1565.191\,nm$). Wavelength shift depends on thermal-expansion coefficient and the thermal-optic coefficient of common single-mode fiber.

Central wavelength and frequency values for FBG temperature sensors are shown in **Table 1**. The wavelength band from 1565.05 to 1565.66 nm are set as a bandwidth of OBPF. The measured spectral curve of the OBPF passband is shown in **Figure 3**. The output of OBPF is connected to one of the optical coupler ports.

For the generation of FOTS data channel signal, the tunable laser diode (LD) with +9 (fiber length 20 km) and 12 dBm (fiber length 40 km) output power, 100 kHz linewidth, 50 dB sidemode suppression ratio (SMSR) is used. LD output is connected with Mach-Zehnder modulator (MZM) with polarization-maintaining PANDA type

Temperature	Wavelength [nm]	Frequency [THz]
−20 °C	1565.66	191.48
26°C	1565.19	191.54
40°C	1565.05	191.55

Table 1.
FBG temperature sensor central wavelength and frequency values.

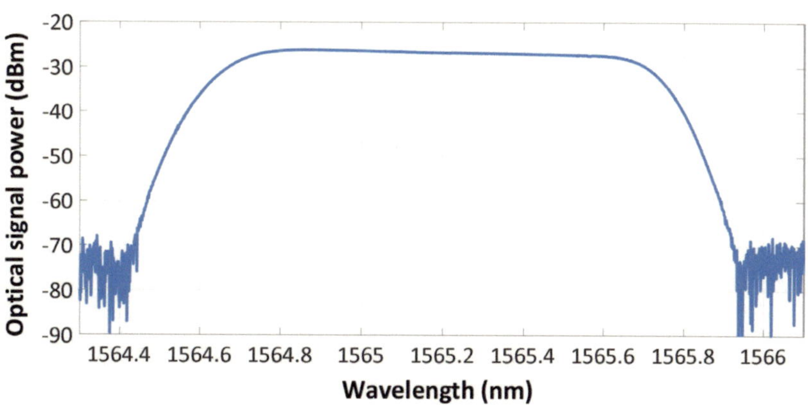

Figure 3.
Measured spectral curve of the OBPF filter passband.

fiber. Electrical data signals are generated by the pattern generator (PPG) (Anritsu, operating bitrate 10 Gbit/s, PRBS 2^{15}–1, signal purity −75 dBc/Hz). PPG data output and electrical RF input of the MZM is connected with proper RF cable. MZM optical output is connected with one of the optical power coupler (OPC) ports. OPC couples signals for FBG sensor and FOTS data channels.

OPC output is connected with the optical circulator (OC) port (1), necessary for separation of the sensor systems optical signal flows (transmitted and reflected). Please see the measured optical circulator insertion loss values in **Table 2**.

The optical circulator port (2) is connected with the optical fiber line and FBG sensor. 20 and 40 km long single-mode optical fiber (SMF-28) spools with insertion loss 4.3 and 8.3 dB are used in these experiments. The optical fiber output is connected with the FBG temperature sensor. FBG sensor structure and operation principle are shown in **Figure 4**.

FBG sensor technology is based on periodical reflection index changes in the fiber core [9–12]. FBG sensor reflects one part of the signal, but another part is transmitted further through the optical fiber. If the object's temperature changes, it shifts transmitted and reflected Bragg wavelength (λ_B), also known as signal central wavelength (see in **Figure 5**). OSA_1 and OSA_2 are used for the analysis of the FBG temperature sensor reflected and transmitted signals.

Bragg wavelength (λ_B) can be described by the following formula (1):

$$\lambda_B = 2 \cdot n_{eff} \cdot \wedge \qquad (1)$$

Where:

Λ – grating period, nm;

n_{eff} – effective group reflection index;

λ_B – Bragg wavelength, nm [10].

FBG sensor temperature is calculated, based on the formula (2):

$$t = t_{ref} + \left(\frac{-\lambda_{ref} + \lambda_{mea}}{\Delta\lambda_{coe}} \right) \qquad (2)$$

Where:

t_{ref} - reference temperature (defined in the sensor specification), °C

λ_{ref} – reference wavelength (defined in the sensor specification), nm

λ_{mea} - measured wavelength value, nm

$\Delta\lambda_{coe}$ – wavelength coefficient, describing wavelength shift, when temperature is changed per 1°C (defined in sensor specification), nm

Temperature sensor parameters used in experiments are listed in **Table 3**.

Direction	Measured insertion loss (dB)
1→2	2
2→3	1
1→3	60

Table 2.
The insertion loss of experimentally used optical circulator.

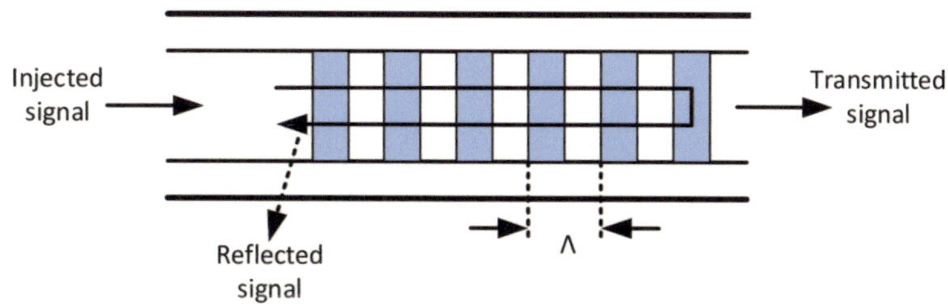

Figure 4.
FBG structure and operation principle.

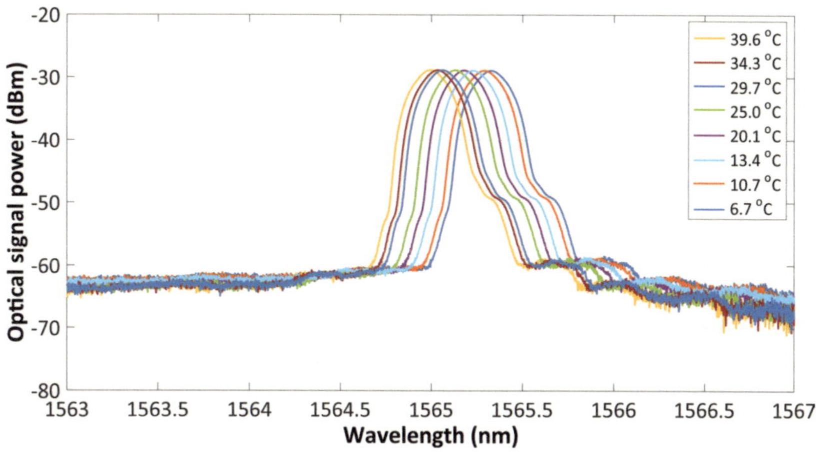

Figure 5.
Measured reflected FBG sensor signal spectrum at different temperatures.

FBG sensor parameter	Value
Reference temperature (t_{ref})	26°C
Reference wavelength (λ_{ref})	1565.191 nm
Wavelength coefficient ($\Delta\lambda_{coe}$)	10.174 pm
Sensor size	3 × 3 × 23 mm
Operation temperature band	−40 °C to +80°C

Table 3.
Parameters of experimentally used FBG temperature sensor.

As we can see in formula 2, and the measured graph (**Figure 6**), the temperature versus wavelength relationship has linear nature.

FBG output is connected with 40-channel (100 GHz channel spacing) arrayed waveguide grating (AWG) flat-top filter with operating wavelength band 1530.334–1561.419 nm (192–195.9 THz). The AWG is used to test the system with different spacing between data and sensor channels. The AWG (with 54 GHz 3-dB and 132 GHz 20-dB bandwidth) 29th and 40th channels are used in experiments, and its parameters are listed in **Table 4**.

AWG output is connected with optical power splitter (20:80%), where 20% are used for signal power monitoring, but 80% are transmitted to photodiode (PD) (with sensitivity (1e-10 BER) = −20 dBm, operation wavelength range = 1280 nm–1580 nm and maximum output voltage = 350 mVp-p) that

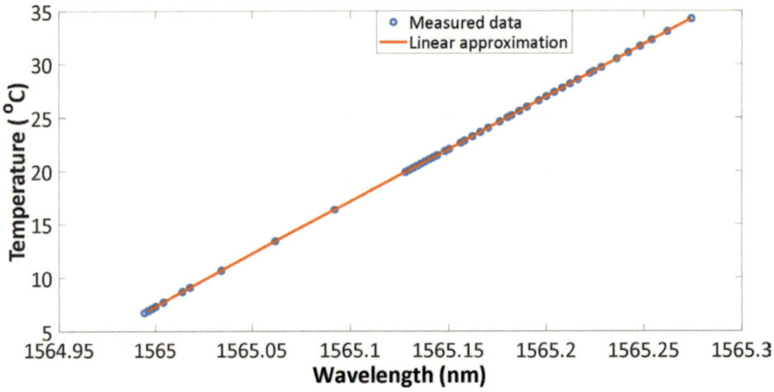

Figure 6.
Temperature versus wavelength relationship for experimentally used FBG temperature sensor.

No. of AWG Channel	Central wavelength [nm]	Frequency [THz]	Attenuation [dB]	
			min	max
29	1552.524	193.1	3.91	4.05
40	1561.419	192.0	4.28	4.41

Table 4.
AWG filter parameters 29th and 40th channel.

converts optical signal to electrical signal. PD output is fed through an RF cable to the eye diagram analyzer (EDA) for data signal quality analysis. For synchronization, the PPG clock signal is transmitted with RF cable to EDA.

Please see the measured FOTS data channel and FBG temperature sensor reflected spectrum in **Figures** 7 and **8**, respectively. FBG temperature sensor reflected signal central wavelength is 1565.1279 nm, and the temperature (calculated with formula (2)) is 19.7°C.

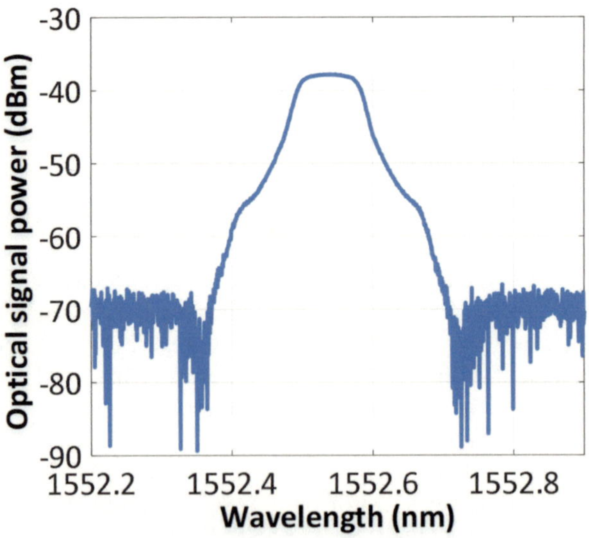

Figure 7.
Measured spectrum of transmission data channel.

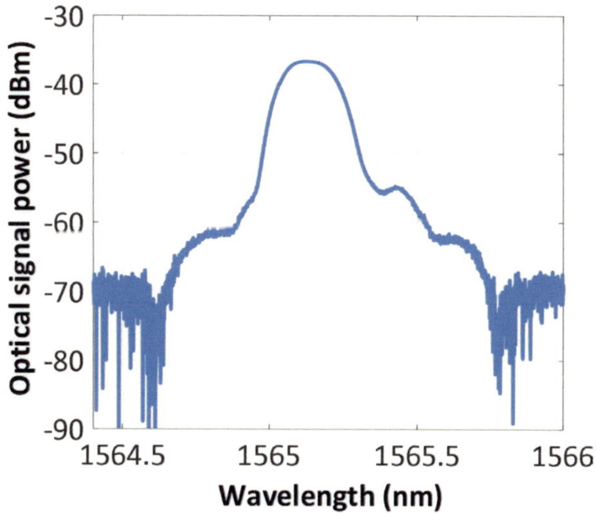

Figure 8.
Measured reflected spectrum of FBG temperature sensor.

Experimentally measured eye diagrams for NRZ-OOK modulated 10 Gbit/s FOTS (after 20 and 40 km long transmission) with and without integrated FBG temperature sensor are shown in **Figure 9**. As we can see in **Figure 9**, the data channel eye diagrams' quality is not degraded by the FBG sensor. Dispersion influence can be observed in eye diagrams (c, d) after 40 km signal transmission, which

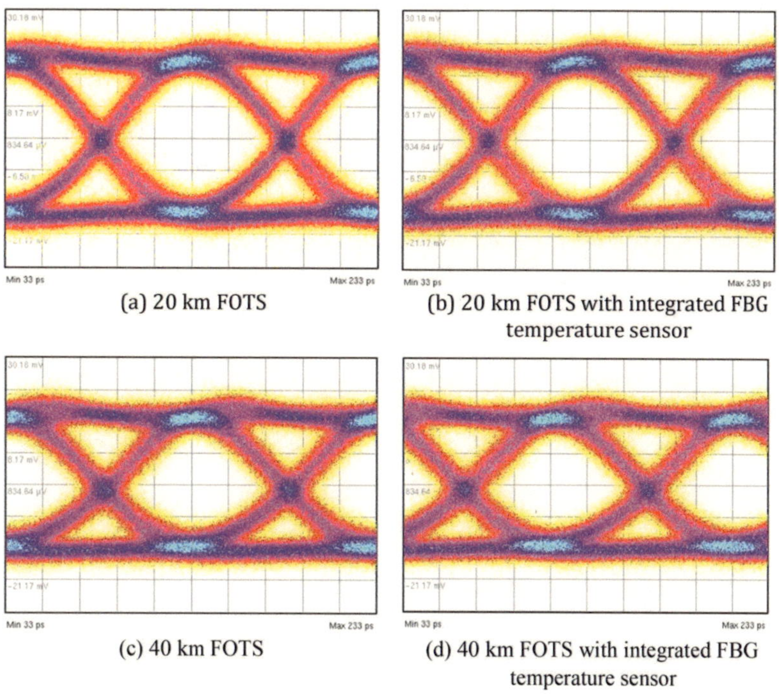

Figure 9.
Comparison of NRZ-OOK modulated 10 Gbit/s FOTS experimental eye diagrams. (a) 20 km FOTS. (b) 20 km FOTS with integrated FBG temperature sensor. (c) 40 km FOTS. (d) 40 km FOTS with integrated FBG temperature sensor.

can be prevented with chromatic dispersion (CD) compensation, such as dispersion compensation fiber (DCF).

3. Evaluation of combined FBG optical sensor network and 32-channel spectrum-sliced wavelength-division-multiplexing passive optical network system in simulation environment

The combined system's simulation setup with the FBG sensor network and SS-WDM PON data channels was developed within RSOFT OptSim software. The simulation setup is shown in **Figure 10**. In this system, only one shared broadband ASE light source is used. ASE spectrum was allocated in the spectral band between from 1533.47 to 1565.50 nm (in frequency band 191.5 THz and 195.5 THz). For SS-WDM PON systems, data transmission channels 1545.7 to 1558.2 nm.

(192.4 THz - 193.95 THz) spectrum was used, whereas 1537.4 to 1545.3 nm (194 THz - 195 THz) spectrum was used for optical FBG sensors network.

ASE light source is connected with an optical power splitter (50%:50%). One of the signal parts is transmitted to AWG MUX for data channel generation, but the second part to OBPF for sensor network. OBPF filtered spectral band from 1537.4 to 1545.3 nm (194 THz - 195 THz) for optical FBG sensor network. 32-channel AWG MUX is used in the setup, that filtered optical signal in the frequency band from 192.4 to 193.95 THz with 50-GHz channel spacing (according to ITU-T G.694.1 recommendation [13]) for 10 Gbit/s NRZ-OOK data channels transmitters. 3-dB bandwidth of each AWG's channel is set to 45 GHz. The data channels bitrate is set to 10 Gbit/s, considering 7% overhead for FEC encoding scheme application resulting in the total bitrate of 10.7 Gbit/s.

Each transmitter block consists of a semiconductor optical amplifier (SOA) to suppress intensity fluctuation noise coming from ASE and electro-absorption modulator (EAM) having immunity to signal polarization state (contrary to MZM). NRZ data signal is generated by data and NRZ component in simulation setup. 32 data channels are coupled with the AWG DEMUX block.

Additionally, DCF is used for dispersion pre-compensation. The transmission line dispersion coefficient is D = 16 ps/nm/km, but the total accumulated dispersion

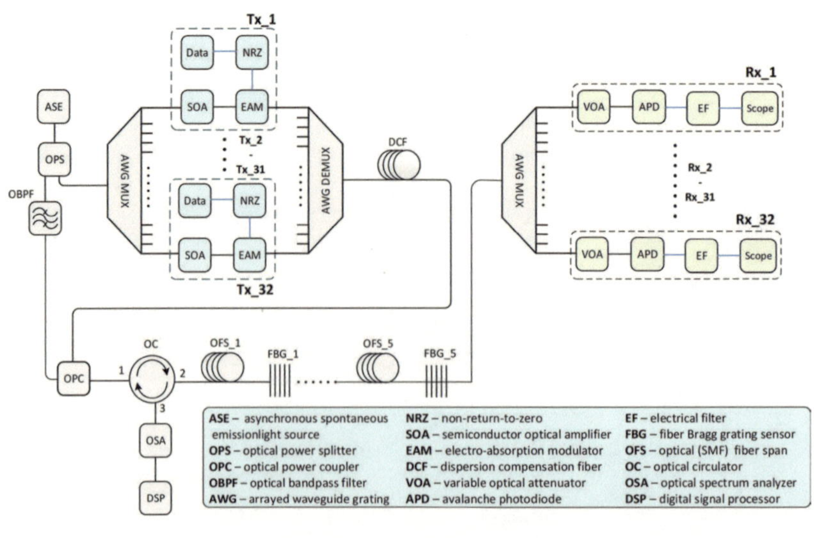

Figure 10.
Simulation setup of combined 32-channel 10 Gbit/s NRZ-OOK modulated SS-WDM-PON system with FBG temperature sensor network.

is 320 ps/nm. DCF fiber dispersion coefficient is −118.5 ps/nm/km, and the attenuation coefficient is α = 0.55 dB/km at 1550 nm reference wavelength. For total dispersion compensation, 2.7 km long DCF is used.

DCF fiber output is connected with one of the OPC input ports, but the second input port is connected with the OBPF output port. The coupled signal is transmitted to OC for separation of the sensor systems signal flows. OC port (2) is connected with 5 optical standard single-mode fiber (ITU-T G.652 recommendation [14]) spans (the length of each span is 4 km, insertion loss is 0.18 dB/km) and 5 FBG temperature sensors network (sensors central wavelength and frequency see in **Table 5**). The spectral band for operation in temperature band −40°C to +80°C (sensor parameters are listed in **Table 3**) is calculated for each sensor. FBG temperature sensor channel spacing is 200 GHz.

The FBG reflected signal from OC port (3) is transmitted to OSA for signal spectrum measurements. Central wavelength or frequency detection, and temperature value calculation are realized in a digital signal processor (DSP).

AWG MUX filtered signal of 32 channels is transmitted to the receiver block. Each receiver block consists of a variable optical attenuator (VOA), avalanche photodiode (APD), electrical filter (EF), and scope components. In this setup, VOA is used for SS-WDM-PON data channels' BER correlation diagram measurements. InGaAs APD (sensitivity set to −20 dBm at the reference BER of 10^{12}) converts optical signal to digital signal. Electrical Bessel low-pass filter 3-dB bandwidth is set to 6 GHz. The received signal quality is analyzed with the scope component, which measures signal eye diagrams and BER value.

FBG sensor's No.	Central wavelength (nm)	Central frequency (THz)
1	1544.53	194.1
2	1542.94	194.3
3	1541.35	194.5
4	1539.77	194.7
5	1538.19	194.9
Channel spacing	1.58	0.2

Table 5.
FBG temperature sensors central wavelength and frequency.

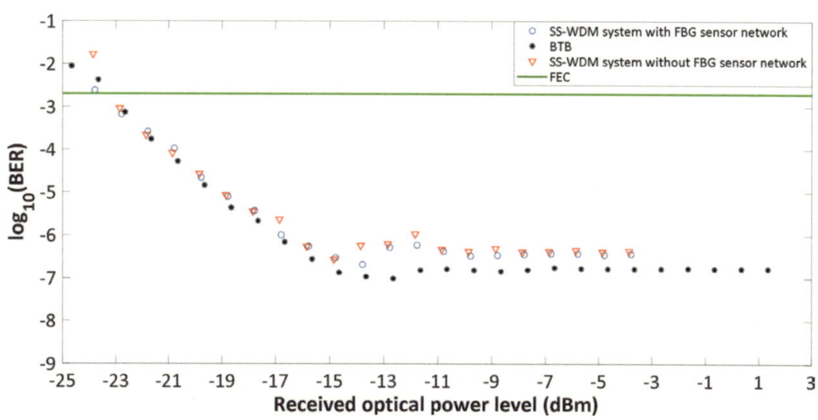

Figure 11.
BER versus received signal power (BER correlation diagrams) for SS-WDM PON system with and without an integrated FBG temperature sensing network.

The 3rd data channel provided the lowest performance of the combined system (SS-WDM-PON data channels and FBG sensor network). Measured BER versus received signal power, known as BER correlation diagram, for SS-WDM-PON system with and without integrated FBG temperature sensing network is shown in **Figure 11**. BER correlation diagram is measured for the worst-performing (in terms of BER) channel of the SS-WDM PON data transmission system.

As we can see in the measured BER versus received signal power (**Figure 11**) graph, the FBG influence on SS-WDM-PON transmission data channels is minimal. FOTS system power reserve is 19.5 dB at the pre-FEC BER level of 2×10^{-3} [15]. The measured BER value of back-to-back (BTB) systems with SS-WDM PON data channels and FBG sensor network is 3.72×10^{-7}, but for back-to-back (BTB) system 1.65×10^{-7}.

Based on the measured results, the calculated power penalty value (compared SS-WDM-PON system with and without integrated FBG temperature sensors network) is 0.5 dB at the pre-FEC BER level 2×10^{-3}.

4. Conclusion

In this research, we successfully demonstrated FBG temperature sensor integration in FOTS experimentally, but the combined system with FBG temperature sensor network and 32-channel spectrum-sliced wavelength-division-multiplexing passive optical network were realized in the simulation environment. The simulation model is based on one shared amplified spontaneous emission source. As shown in **Figures 9** and **11**, FBG sensors do not degrade data channel signal quality.

The measured BER value for the worst channel of a back-to-back (BTB) systems with SS-WDM-PON data channels and FBG sensor network is 3.72×10^{-7}, but for BTB system 1.65×10^{-7}. The FOTS systems power reserve is 19.5 dB at the pre-FEC BER level of 2×10^{-3}.

Acknowledgements

This work is supported by the European Regional Development Fund project No. 1.1.1.3/18/A/001, PVS 3912.6.2.

Conflict of interest

The authors declare no conflict of interest.

Data availability

The data used to support the findings of this study are available from the first author upon request.

Author details

Janis Braunfelds[1*], Sandis Spolitis[1], Jurgis Porins[2] and Vjaceslavs Bobrovs[2]

1 Riga Technical University, Communication Technologies Research Center, Riga, Latvia

2 Riga Technical University, Institute of Telecommunications, Riga, Latvia

*Address all correspondence to: janis.braunfelds@rtu.lv

IntechOpen

© 2020 The Author(s). Licensee IntechOpen. This chapter is distributed under the terms of the Creative Commons Attribution License (http://creativecommons.org/licenses/by/3.0), which permits unrestricted use, distribution, and reproduction in any medium, provided the original work is properly cited.

References

[1] Taylor & Francis Group. Optical Fiber Sensors – Advances Techniques and Applications. CRC Press. 2015. pp. 545. DOI: 10.1201/b18074.

[2] Fidanboylu K., Efendioğlu H. S.b. Fiber optic sensors and their applications. In: 5th International Advanced Technologies Symposium (IATS'09); 13-15 May 2009; Karabuk, Turkey. pp. 7.

[3] Senkans U., Braunfelds J., Lyashuk I., Porins J., Spolitis S., Bobrovs V. "Research on FBG-Based Sensor Networks and Their Coexistence with Fiber Optical Transmission Systems". Hindawi. Journal of Sensors. 2019. 6459387.DOI: 10.1155/2019/6459387.

[4] Marcelo M. Werneck, Regina C. S. B. Allil, Bessie A. Ribeiro and Fábio V. B. de Nazaré. A guide to Fiber Bragg Grating Sensors. In: Christian Cuadrado-Laborde editors. Current Trends in Short- and Long-period Fiber Gratings. IntechOpen. 2013, pp. 24. DOI: DOI: 10.5772/54682.

[5] Braunfelds J., Senkans U., Lyashuk I., Porins J., Spolitis S., Bobrovs V. Unified Multi-channel Spectrum-sliced WDM-PON Transmission System with Embedded FBG Sensors Network. In: (2019) Progress in Electromagnetics Research Symposium; 17-20 June 2019; Rome, Italy; 2019. pp. 3327-3333.

[6] Senkans U., Braunfelds J., Lyashuk I., Porins J., Spolitis S., Haritonovs V., Bobrovs V. FBG sensors network embedded in spectrum-sliced WDM-PON transmission system operating on single shared broadband light source. In: 2019 Photonics and Electromagnetics Research Symposium - Fall, PIERS - Fall 2019 – Proceedings; 17-20 Dec. 2019; Xiamen, China; 2019. pp. 1632-1639.

[7] Campanella C.E., Cuccovillo A., Campanella C., Yurt A., Passaro V.M.N. "Fibre Bragg Grating based strain sensors: Review of technology and applications". MDPI. Sensors. 2018. 18 (9). 3115. DOI: 10.3390/s18093115.

[8] Laffont G., Cotillard R., Roussel N., Desmarchelier R., Rougeault S. "Temperature resistant fiber bragg gratings for on-line and structural health monitoring of the next-generation of nuclear reactors". MDPI. Sensors. 2018. 18 (6). 1791. DOI: 10.3390/s18061791.

[9] Laferrière J., Lietaert G., Taws R., Wolszczak S., "Reference Guide fo Fiber Optic Testing" JDSU. pp. 172. 2012.

[10] Qiao X., Shao Z., Bao W., Rong Q., "Fiber Bragg Grating Sensors for the Oil Industry". MDPI. Sensors. 2017. 17 (3). 429. DOI: 10.3390/s17030429.

[11] Zhang M., Xing Y., Zhang Z., Chen Q. "Design and Experiment of FBG-Based Icing Monitoring on Overhead Transmission Lines with an Improvement Trial for Windy Weather". MDPI. Sensors. vol. 14, no. 12. DOI: 10.3390/s141223954.

[12] Qiao X., Shao Z., Bao W., Rong Q. "Fiber Bragg Grating Sensors for the Oil Industry". MDPI. Sensors. 2017. 17. DOI: 10.3390/s17030429.

[13] ITU-T Recommendation G.694.1, Spectral grids for WDM applications: DWDM frequency grid, International Telecommunication Union, Telecommunication standardization sector of ITU, 2012, pp. 1-7.

[14] ITU-T Recommendation G.652, Characteristics of a single-mode optical fibre and cable, Telecommunication Union, Telecommunication standardization sector of ITU, 2017, pp. 1-28.

[15] Spolitis S., Bobrovs V., Parts R., Ivanovs G. Extended reach 32-channel dense spectrum-sliced optical access system, In: Progress in Electromagnetic Research Symposium (PIERS); 8-11 Aug. 2016. Shanghai, China; 2016. pp. 3764-3767.

Chapter 6

Analysis of Energy Transfer among Stokes Components in Raman Fiber Lasers

Lelio de la Cruz-May, Efraín Mejía Beltrán, Olena Benavides and Rafael Sanchez Lara

Abstract

This work presents a methodology to estimate the pumping power required for the first Stokes to reach its maximum stored energy level, before it generates the next Stokes. These estimates are achieved by experimentally measuring the critical power and the relationship between the pumping power (P_{P0}) and the small signal of the stimulated Raman spread (P_{F0}). For our study we used 1 km of 1060-XP fiber, experimentally obtaining P_{cr} = 6.693 W, P_{F0}/P_{P0} = 6.759x10^{-6}. With these experimental data, the pump power required for the first Stokes to reach its maximum stored energy level was 13.39 W, and the stored energy in the first Stokes was 9.88 W. It is important to note that the Raman threshold $\ln(P_{P0}/P_{F0})$ = 11.9 is smaller than the initially reported ~16.

Keywords: fiber optics, Raman scattering, Raman fiber lasers

1. Introduction

Stimulated Raman Scattering (SRS) in optical fibers has been an intensified subject of studies during the last two decades; at the beginning it was investigated in non-doped optical fibers but during the last decade the hybrid rare-earth-doped combined with the pure Raman effect has been developed and attracted much attention [1]. The SRS is a phenomenon responsible of a new generation of fiber lasers and Raman amplifiers because the high brightness of the confined powers traveling and interacting along optical fibers greatly facilitate the Raman scattering (RS) based amplification. Raman scattering occurs when a monochromatic light beam propagates along an optical fiber, whereas most of the beam power is transmitted without change, a small quantity in the order of 10^{-6} per kilometer of fiber is scattered isotropically with a new frequency [2]. At high coupled pump powers, the stimulated version of this phenomenon, the SRS, occurs and the portion of this frequency-shifted scattered light that travels through the fiber core becomes amplified and demands energy to feed itself from the pump power in order be amplified. This amplification is very efficient such that most of the traveling pump power can be transferred to this Stokes component. Therefore, the SRS is the platform for the development of Raman fiber lasers and amplifiers. As the Stokes component grows very remarkably, the pump power decreases until it is unable to continue transferring energy. In a sufficiently pumped optical fiber such that the pump is practically

IntechOpen

extinguished during this energy transfer, now this Stokes can be high enough to produce the following Stokes by the same SRS-process; again, a strong enough second Stokes can produce a third one and so on. Using the historically well developed laser operating around 1064 nm as a pump source for this type of Raman-based lasers it is possible to obtain Stokes components that cover the 1.1–1.7 μm region that is of great importance for several applications such as: optical fiber communications (the internet platform), material processing (cutting, soldering, ablating, etc.), laser spectroscopy and medicine [3]. Equations for the behavior of the Raman amplification have been described in [4, 5] and their solutions have been proposed in [4, 6]. Despite these advances with the propagation equations describing the energy transfer among stokes components in Raman amplification, there is a lack of simple methods that allow the exact estimation of some important parameters such as: the Raman threshold power, the maximum pumping power necessary for the power stored in the Stokes signal to be maximized, the Raman gain coefficient and the power required to obtain any N-order of Stokes.

This chapter describes the maximum energy that can be stored in a given Stokes line as after such maximum storage it starts producing the next Stokes. From the existing physical–mathematical theories, a formula has been obtained by us and has been verified with experimental results.

2. Theory

In order to find a mathematical relationship that predicts the maximum power stored in the Stokes signal is necessary to solve the equations that describe the forward propagation of the pumping power and Stokes line in a single mode fiber given by [4, 7, 8];

$$dP_P/dz = -\alpha_P P_P - (\upsilon_P/\nu_S)\left(g_r/A_{eff}\right)P_P P_F \tag{1}$$

$$dP_F/dz = -\alpha_S P_F + \left(g_r/A_{eff}\right)P_P P_F \tag{2}$$

where P_P and P_F, ν_S and ν_P, α_P and α_S are powers, frequencies and loss coefficients of pump power and Stokes signal, respectively. A_{eff} is effective area, and g_r is the Raman gain coefficient of the Stokes signal.

Assuming that $\alpha_p \approx \alpha_s$ it is possible to minimize the errors by establishing $\alpha = (\alpha_p + \alpha_s)/2$ given that the attenuation curve of the silica fiber is approximately constant in the region covering pump and first Stokes signals. An analytical solution of the equations that govern the SRS can be obtained by the following procedure:

A. First dividing the Eq. (1) between Eq. (2),

$$\frac{dzdP_P}{dzdP_F} = \frac{\left(-\alpha - \frac{\upsilon_P}{\upsilon_s}\frac{gr}{A_{eff}}P_F\right)P_P}{\left(-\alpha + \frac{gr}{A_{eff}}P_P\right)P_F} \tag{3}$$

Where the term dz is canceled, and consequently we proceed to the separation of variables, to obtain the following mathematical expression:

$$\left(-\alpha + \frac{gr}{A_{eff}}P_P\right)\frac{dP_P}{P_P} = \left(-\alpha - \frac{\upsilon_p}{\upsilon_s}\frac{gr}{A_{eff}}P_F\right)\frac{dP_F}{P_F} \tag{4}$$

Solving the differential equation by the separable equation method, we obtain the following expression:

$$\ln\left[\left(\frac{P_{FL}}{P_{F0}}\right)\left(\frac{P_{P0}}{P_{PL}}\right)\right]=\left(\frac{g_r}{A_{eff}}\right)\left(\frac{\upsilon_P}{\nu_S}\right)\frac{1}{\alpha}\left[\left(\left(\frac{\upsilon_S}{\nu_P}\right)P_{P0}+P_{F0}\right)-\left(\left(\frac{\upsilon_S}{\nu_P}\right)P_{PL}+P_{FL}\right)\right] \tag{5}$$

Where the parameters represent: P_{PL} output pump power, P_{PO} input pump power, P_{FL} output power Stokes and P_{FO} initial power Stokes, and evaluated in all the fiber from $z = 0$ to $z = L$.

B. Adding the Eqs. (1) and (2).

First, we will multiply the Eq. (2) by (ν_p/ν_s), and consequently, the algebraic operation results:

$$\frac{d}{dz}\left(P_P+\frac{\upsilon_p}{\upsilon_s}P_F\right)=-\alpha\left(P_P+\frac{\upsilon_p}{\upsilon_s}P_F\right) \tag{6}$$

Performing algebraic operation and integrating by separable variable, we obtain:

$$\left(\left(\frac{\nu_S}{\nu_P}\right)P_{PL}+P_{FL}\right)=\left(\left(\frac{\nu_S}{\nu_P}\right)P_{P0}+P_{F0}\right)e^{-\alpha L} \tag{7}$$

where P_{P0} and P_{PL} are the input and output powers, P_{F0} and P_{FL} are initial and output Stokes signal. L is the fiber length.

On the other hand, for a 1-Km of any telecom optical fiber, approximately a few millionths of the input pump power (P_{PO}) becomes available at the output as RS (spontaneous, linear) Stokes signal (P_{FO}), this is practically the signal that initiates the amplification process once the stimulated process takes place. Therefore, the term $((\nu_S/\nu_P)P_{PO} + P_{FO}) \approx (\nu_S/\nu_P)P_{PO}$ [2]. Substituting Eq. (7) in Eq. (5), it is obtained a general solution to differential Eqs. (1) and (2), given by [6]:

$$\ln\left(\frac{P_F}{P_p}\frac{P_{P0}}{P_{F0}}\right)=\frac{g_r}{A_{eff}}P_{P0}L_{eff} \tag{8}$$

where $L_{\text{eff}} = [(1\text{-}\exp(-\alpha L))/\alpha]$ represents the interaction length of the pumping and the Stokes signal. Eq. (5) provides information on parameters such as the Raman gain, the powers of the Pump and Stokes signals at the output end of the fiber.

3. Raman threshold

Raman threshold is a very important parameter since above this power the energy transfer from pump to the first Stokes becomes highly efficient (**Table 1**). In optical amplification, it is important to operate above threshold; on the contrary, not reaching threshold becomes crucial in multichannel communications systems as non-linearly produced Stokes-copies of any channel can interfere with neighboring channels. [2].

Figure 1 presents the output powers of: (a) Residual pump (black line with squares) and (b) Stokes signal at 1117 nm (red line with circles) at the output of

Fiber span [km]	L_{eff} [km]	P^{1st}_{Max} [W]	P_1^{Sto} [W]
1	0.846	13.39	9.88
2	1.443	7.48	3.9
5	2.371	4.77	0.88

Table 1.
Three estimates of maximum stokes power for 1060-XP fiber.

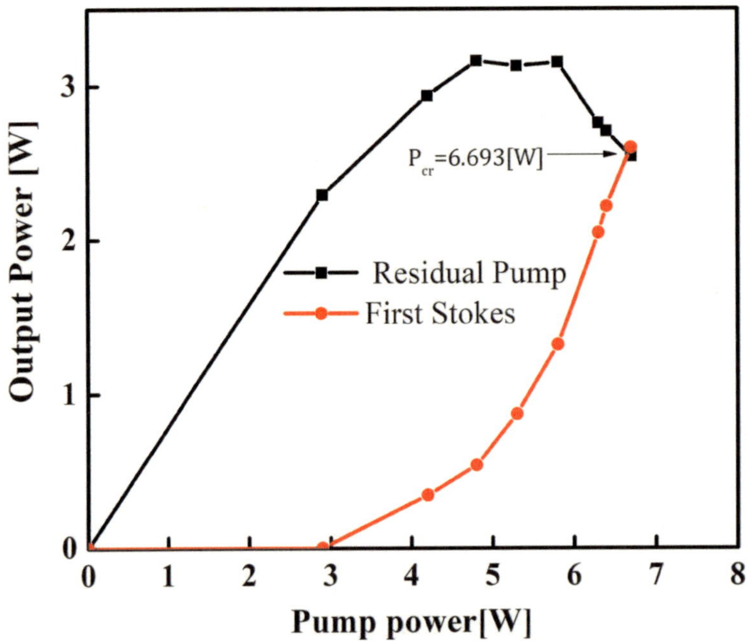

Figure 1.
Residual-pump and stokes powers delivered at the output of an optical fiber as functions of the pump power coupled at the input.

1-km 1060-XP optical fiber as a functions of 1064-nm pump power. Observe that the aforementioned threshold pump was around 2.9 W. There is another important parameter in which the residual and Stokes powers are equal, this is called the critical (pump) power that occurs at 6.693 W. These curves were obtained by splicing the 1060-XP optical fiber to an Ytterbium-based 1064-nm fiber laser without forming a laser cavity; i.e., in a free-running configuration.

At critical power ($P_P = P_F$) from Eq. (8) we have:

$$\ln\left(\frac{P_{P0}}{P_{F0}}\right) = \frac{g_r}{A_{eff}} P_{Pcr} L_{eff} \tag{9}$$

Taking into account that for standard fibers of 1-km length P_{F0} is in the order of $P_{P0}x10^{-6}$; under this assumption, Eq. (9) usually approximates:

$$16 \approx \frac{g_r}{A_{eff}} P_{cr} L_{eff} \tag{10}$$

where P_{P0} is equal to the critical power P_{cr}. This relationship has been previously reported [4] and used in several research works presenting an acceptable accuracy.

Modern specialty optical fibers are designed for purposes other than optical communications and are usually doped with materials such as P_2O_5, GeO_2 and B_2O_4

to improve some of their properties for different applications; for such special fibers the Eq. (10) is not a good approximation, in this sense, the relationship P_{P0}/P_{F0} must be determined experimentally, as it is the case in this work. This ratio is unique for each optical fiber, that is, each optical fiber has its own response to the stimulated Raman scattering [9].

4. Raman gain

The Raman gain coefficient g_r describes how the Stokes power grows as the pump power is transferred through the stimulated Raman scattering; in our case, given that the optical fiber under test has common characteristics, Eq. (9) is useful to calculate the Raman gain. Note that the effective fiber length is a constant that depends on the loss coefficient and the fiber length.

An analysis of the behavior below threshold, at threshold and above threshold is presented by the corresponding spectra on **Figure 2**. These spectra exactly correspond to the same experiment corresponding to **Figure 1**.

Observe that at 2.9-W pump power, the Stokes power has −68.5 dBm, this is the maximum RS level P_{F0} (the figure presents an error, instead of P_{P0} it should be P_{F0}). The evolution of this signal is presented until the SRS signal is evident at 3.3-W pump [9].

At any pump power, at the fiber output the relation between the output coupled pump power P_P and the produced Raman signal (P_{F0}) may be quantified by:

$$\frac{P_P}{P_{FO}} = 10^{\frac{\Delta Power(dBm)}{10}} \tag{11}$$

where $\Delta Power = P_P - P_{F0}$. When the spontaneous Raman just appears at the output, practically, the pump signal only suffers linear attenuation and thus the relationship between the residual pump power $P_P(L)$ and the coupled pump power P_{P0} is:

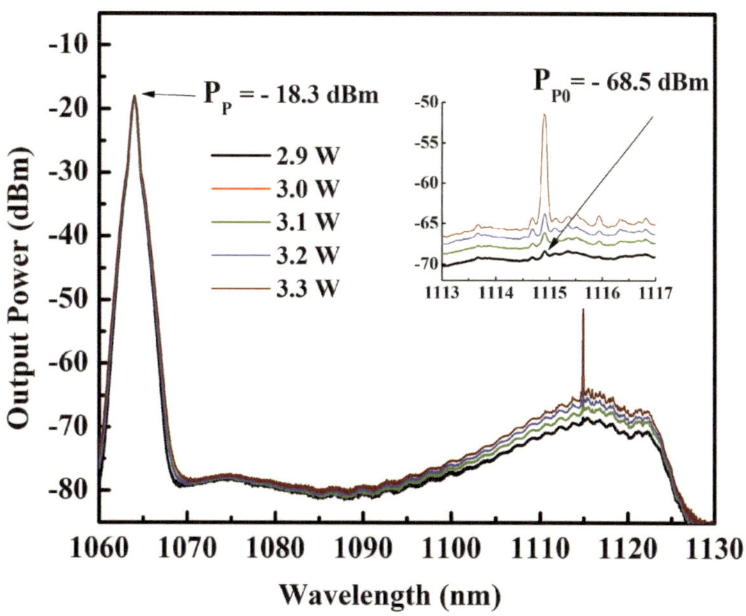

Figure 2.
Output signals of the 1-km 1060-XP optical fiber.

$$P_P = P_{P_0} e^{-\alpha L} \tag{12}$$

Now, substituting Eq. (7) in (6), it is possible to numerically establish a relationship for P_{PO}/P_{FO} given by:

$$\frac{P_{PO}}{P_{FO}} = 10^{\frac{\Delta Power(dBm)}{10}} e^{\alpha L} \tag{13}$$

Note that the term on the left side of Eq. (13) is constant, and it depends on the physical characteristics of the optical fiber and its response to spontaneous RS [8]. And it can be evaluated just when the spontaneous Raman Scattering becomes stimulated Raman scattering, then this relationship can be obtained experimentally from **Figure 2**.

Then using this numerical result, one obtains the Raman gain efficiency by rewriting Eq. (9) as:

$$g_R = \frac{g_r}{A_{eff}} = ln\left(\frac{P_{P0}}{P_{F0}}\right)\frac{1}{P_{cr}L_{eff}} \tag{14}$$

Figure 2 shows the spectral composition of the signal delivered by te 1-km 1060-XP optical fiber. Note that, −18.3 dBm level corresponds to the residual pump $P_P(L)$. At the lowest pump power P_{PO}, only spontaneous RS occurs with signal from 1090 to 1130 nm, its maximum level is around -68.5 ± 1 dBm. Taking this level as P_{FO}, the difference is ΔPower = 50.2 ± 1 dBm. The fiber loss is 1.5 dB/Km at 1064 nm according with the technical specification provided by de manufacturer. Substituting these data on the Eq. (13) yields P_{PO}/P_{FO} = 147,945.506 and $\ln(P_{PO}/P_{FO})$ = 11.9, hence $P_{FO} = 6.759\text{x}10^{-6}P_{PO}$. These numbers are of great importance, instead of obtaining the 16-number of Eq. (10), we have obtained 11.9, meaning that this is not considered a standard telecommunications fiber, this is true since it was designed for operating in the 1060-nm region; also note that the RS signal produced by 1-km of this fiber was $6.76\text{x}10^{-6}$ multiplied by the pump power. Everything is within the values expected.

Using the critical power P_{cr} = 6.693 W from **Figure 1**, and calculating the effective length L_{eff} = 0.846 km one may use the Eq. (14) to obtain the Raman gain efficiency $g_R = 2.1[\text{W}^{-1}\ \text{km}^{-1}]$; very important value to be used to model the propagation of pump and stokes in a fiber Raman laser system using Eqs. (1) and (2).

5. Maximum power stored in first stokes

Once the Raman threshold is reached, first Stokes grows quickly until the pump power is unable to continue transferring power, and this energy transfer is described by Eq. (8). When the energy transfer ceases, the first Stokes reaches its maximum value, denoted by P^{1st}_{Max}, and precisely the first Stokes is capable of generating the second Stokes. We have previously known that approximately six millionth of pump power is transferred in spontaneous RS, that is, $P_{FO}/P_{PO} = 6.759\text{x}10^{-6}$. We propose that the pump power ceases to transfer energy to the first Stokes when it reaches a value of six millionth multiplied by the threshold power. In this sense, and to estimate an approximate value to produce the next Stokes, we propose this be similar to $P_{FO}/P_{PO} = P_P/P_F = 1/\varepsilon$, in such a way that Eq. (8) becomes;

$$\text{P}^{\text{1st}}_{\text{Max}} = 2\ln(\varepsilon)\frac{A_{eff}}{g_r}\frac{1}{L_{eff}} \tag{15}$$

Eq. (15) represents the pumping power that must be used for the first Stokes to reach its maximum energy level. Substituting Eq. (15) in Eq. (7), and neglecting the terms P_{F0} and P_{PL}, because their values are very small, consequently, an equation is obtained for the power stored in the first Stokes, given by:

$$P_1^{Sto} = \frac{\upsilon_S}{\nu_P} 2\ln(\varepsilon) \frac{A_{eff}}{g_r} \frac{1}{L_{eff}} e^{-\alpha L} \tag{16}$$

Taking previous data for the 1060-XP fiber, $\ln(\varepsilon) = \ln(P_{P0}/P_{F0}) = 11.9$, $g_R = 2.1$ $[W^{-1}\,km^{-1}]$ and 1.5 dB/Km. On the other hand, using $\lambda s = 1120$ nm and $\lambda p = 1064$ nm, we can estimate the maximum pump power and the maximum power stored in the first Stokes, for different fiber span.

In general, the pumping power for storing the maximum energy in N-order Stokes is given by;

$$P_{Max}^{N} = 2\ln(\varepsilon) \frac{A_{eff}}{g_r} \frac{1}{L_{eff}} \left[1 + \frac{\lambda_P}{\lambda_s} + \frac{\lambda_s}{\lambda_{s2}} + \ldots + \frac{\lambda_{N-1}}{\lambda_N}\right] \tag{17}$$

With this equation, we obtain approximately the maximum power stored in the N-order Stokes before generating the N + 1-order Stokes. This equation is versatile because the parameters involved are simple to calculate. To date, the Eq. (17) at least of our knowledge has not been reported.

And finally, the ratio between the maximum power stored and coupled pump power for the first Stokes is given by;

$$\frac{P_1^{Sto}}{P_{Max}^{1st}} = \frac{\lambda_P}{\lambda_s} e^{-\alpha L} \tag{18}$$

This equation depends on the attenuation and the fiber length, and describes the optical conversion efficiency for a particular fiber length (L) and fiber attenuation (α). If $L = 0$, the Eq. (18) gives the maximum energy conversion efficiency or quantum efficiency.

In this sense, for a longer optical fiber the energy conversion efficiency is lower. For example, for 5 km of 1060-XP fiber this value can be ~17%. This implies that the fibers for the design of Raman lasers and Raman amplifiers must be of a length that allows an acceptable conversion efficiency [10, 11].

6. Conclusions

An analytical equation was developed to predict the pumping power necessary to store the maximum energy in the first Stokes, this is a powerful tool for numerical simulations, which can provide specific data for the design of Raman fiber lasers. Consequently, an analytical equation is deduced that can predict the stored power in the following Stokes orders, that is, progressively, the second, the third, etc.

For a 1 km fiber span, 13.39 W of pumping power is required to get the first Stokes to reach its maximum stored energy level of 9.88 W, while for 5-km of fiber span a pump power of 4.77 W is required to obtain 0.88 W. It is important to mention that the experimental calculation of $\ln(P_{P0}/P_{F0})$ let us conclude that the fiber is not intended for telecommunications as the ~16 number is not satisfied.

Finally, it is important to note that commercial single-mode fiber lasers normally operate at pump powers of around 10 W, therefore, an estimate such as the one made in this work is fundamental.

Author details

Lelio de la Cruz-May[1*], Efraín Mejía Beltrán[2], Olena Benavides[1] and Rafael Sanchez Lara[1]

1 Universidad Autónoma del Carmen, Facultad de Ingeniería, Ciudad del Carmen, Campeche, México

2 Centro de Investigaciones en Óptica, León, Guanajuato, México

*Address all correspondence to: ldelacruz@pampano.unacar.mx

IntechOpen

© 2020 The Author(s). Licensee IntechOpen. This chapter is distributed under the terms of the Creative Commons Attribution License (http://creativecommons.org/licenses/by/3.0), which permits unrestricted use, distribution, and reproduction in any medium, provided the original work is properly cited.

References

[1] L. de la Cruz-May, E.B. Mejia: Raman Fiber Laser Improvement by Using Yb3+−Doped Fiber. Laser Phys. (2009); 19: 1017–1020 . https://link.springer.com/article/10.1134/S1054660X09050235

[2] G. P. Agrawal, Nonlinear Fiber Optics, 5th Ed. Oxford: Elsevier Academic Press; 2013. 295 p. http://dx.doi.org/10.1016/B978-0-12-397023-7.00008-5

[3] A. Kurkov, E. Dianov, V. Paramonov, A. Gur'yanov, et. al.: High-power fibre Raman lasers emitting in the 1.22–1.34 µm range. Quantum Electron. (2006); 30: 791–793. https://doi.org/10.1070/QE2000v030n09ABEH001806

[4] R. Smith: Optical Power Handling Capacity of Low Loss Optical Fibers as Determined by Stimulated Raman and Brillouin Scattering. Appl. Opt. (1972); 11: 2489–2494. https://doi.org/10.1364/AO.11.002489

[5] J. AuYeung, A. Yariv: Theory of cw Raman oscillation in optical fibers. J. Opt. Soc. Am. (1979); 69: 803–807. https://doi.org/10.1364/JOSA.69.000803

[6] L. de la Cruz-May, J. A. Alvarez-Chavez, E. B. Mejia, A. Flores-Gil, F. Mendez-Martinez, S. Wabnitz: Raman threshold for nth-order cascade Raman amplification. Opt. Fiber Technol. (2011); 17: 214–217. https://doi.org/10.1016/j.yofte.2011.02.002

[7] J. Bromage: Raman Amplification for Fiber Communications Systems. J. Lightwave Technol. (2004); 22: 79–93. DOI:10.1109/JLT.2003.822828

[8] G. Pei-Juan, N. Cao-Jian, Y. Tian-log, and S. Hai-Zheng: Stimulated Raman scattering up to 10 orders in an optical fiber. Appl. Phys. (1981); 24: 303–306. https://doi.org/10.1007/BF00899726

[9] L. de la Cruz-May, E. B. Mejia, O. Benavides, J. Vasquez Jimenez, J. Castro-Chacon, and M. May-Alarcón: Novel Technique for Obtaining the Raman Gain Efficiency of Silica Fibers. IEEE Photonics Journal. (2013); 5: 6100305–6100305. doi: 10.1109/JPHOT.2013.2271900

[10] S. Huang, Y. Feng, A. Shirakata and K.I. Ueda: Generation of 10.5 W, 1178 nm Laser Based on Phosphosilicate Raman Fiber Laser. Jpn. J. Appl. Phys. (2003); 42: L1439. https://doi.org/10.1143/JJAP.42.L1439

[11] Codemard, C., J. Ji, J. K. Sahu and J. Nilsson. 100-W CW cladding-pumped Raman fiber laser at 1120 nm. Proceedings SPIE 7580, Fiber Lasers VII: Technology, Systems, and Applications. 17 February 2010. doi: 10.1117/12.845606